WEATHER OF THE PACIFIC COAST

WEATHER OF THE PACIFIC COAST

WASHINGTON
OREGON
BRITISH COLUMBIA

Walter Rue

The Writing Works, Inc.
Mercer Island, Washington

Gordon Soules Book Publishers
Vancouver, B.C. Canada

Library of Congress Cataloging in Publication Data

Rue, Walter D., 1905-
Weather of the Pacific Coast.

Includes index.
1. Washington (State)—Climate. 2. Oregon—Climate. 3. British Columbia—Climate.
I. Title.
QC984.W2R83 551.6'9'795 78-13389
ISBN 0-916076-21-0

Edited by Wendy Garner
Cover and illustrations by Bob McCausland

Manufactured in the United States
Published by The Writing Works, Inc.
7438 S.E. 40th Street
Mercer Island, WA 98040
ISBN: 0-916076-21-0 (U.S.A.)
Library of Congress Catalog Card Number: 78-13389

Published in Canada by Gordon Soules Book Publishers
1118-355 Burrard Street
Vancouver, B.C., Canada V6C 2G8
ISBN: 0-919574-27-0 (Canada)

To friends in the National Weather Service
who unlocked many meteorological mysteries for me,
and to my associates in journalism.

Contents

If wishing could but make it so
The balmy days would never go;
The sun would shine from morn till night,
And dispositions would be bright.

The rains would come just now and then,
And mostly in the nighttime when
Our flowers, shrubs and trees grown tall,
Could use the moisture best of all.

A "high" is needed, bright and clear,
Above us in the atmosphere
To ward away the lowly "lows"
That bring us all our weather woes.

Foreword

I CAN'T remember when I did not know Walt Rue around the cluttered city room of the *Seattle Post-Intelligencer* where we toiled for more years than I care to count. We both landed on the P-I staff in the Roaring Twenties, I as a cub reporter assigned to day police and Walt as a young copy boy fresh out of school.

Walt handled the humble but essential job of copy boy with such quiet competence that in due course he was promoted to the reporting staff and became a quietly competent reporter with a remarkable command of precise English.

Our entry into World War II brought profound changes in both our lives. Walt went off to war, and I was dragged kicking and screaming into the job of managing editor under the stern and demanding eye of Mr. Hearst who owned the paper. (I didn't think I would last six months.)

The next thing we stay-at-homes knew Walt was in Canada's Far North, assigned to the 16th Weather Squadron of the Army Air Corps. The squadron had established numerous observing stations throughout Canada, and in Alaska, and some of those stations were in the continent's coldest hinterlands near the Arctic.

After graduating from the weather observers' school at Chanute Field, Illinois, Walt was sent to one of those northern outposts with three other observers and three members of the Signal Corps. The station was in northern Saskatchewan.

You would think that duty in the Far North would have cured him of any desire to pursue the subject beyond the call of duty. But he had been fatally bitten by a meteorological bug.

His Army career, following graduation from the school in Illinois, included duty at two observing stations. Then, to his surprise, he was transferred to Squadron headquarters in Great Falls, Montana. The Colonel in command had learned through the Army grapevine that Walt was a newspaperman, and the Colonel needed someone to write the Squadron history.

So Walt became a historian. He also diligently pursued the subject of meteorology, which had become a hobby. He had found his niche in life.

When Walt returned to the P-I staff after the war, someone around the shop had the brilliant idea of making him a full-fledged weather reporter. Thus the *Post-Intelligencer* was unique among all the metropolitan dailies of the United States in having a regular weather column by a staff reporter who had received extensive training in meteorology. And the wonder is why nobody had thought of such a thing before and why it's still uncommon in the newspaper business.

For what topic is of more universal interest than the weather? It governs our lives as does no other condition of living on this planet. It's the first order of conversation when we greet friends or casual acquaintances on the street or in the shops. It sets our mood from day to day and from hour to hour; we are apt to be depressed on a gray rainy day and cheerful and optimistic when the sun is shining brightly. It dictates the prices we pay for the food that sustains our lives. In its many violent forms it is the greatest killer of all.

Walt wrote of these things in the *Post-Intelligencer,* and they were exciting reading: Seattle's big snow of February, 1916; the great storm of October 12, 1962, perhaps the worst windstorm ever experienced on the Pacific Coast in historical times; the 1910 Wellington train disaster in the Cascades; that winter of 1867-68 when the lower Columbia River froze solidly from bank to bank; Seattle's only blizzard in January, 1950.

But dearer to Walt's heart than these infrequently violent manifestations of our Northwest weather are those more normal climatic conditions which make this corner of the United States one of the finest places for human habitation in the world: the beauty of our lakes and mountains, our forest and our beaches, our gardens and our rivers. He writes of these things also in loving detail.

His weather columns in the *Post-Intelligencer* brought him a number of awards including one from the American Meteorological Society for "outstanding work in writing feature articles on the subject of weather and climate." Others were from the Seattle Historical Society, the National Weather Service and Sigma Delta Chi, the national journalism society.

Forecast for today: the chance that you will be fascinated by the pages that follow, 90 percent or more.

ED STONE

1

Washington— A State of Contrasts

WASHINGTON is a state of unusual climatic contrasts. Parts of the state get more rain than any other section of the nation. Other places are so dry they qualify as desert country. Extremes of heat and cold are unusual on the west side of the Cascade Mountains and commonplace in some areas on the east side of the mountain range.

From Drenched to Dry

Wynoochee Oxbow, at an elevation of 600 feet in the Wynoochee River Valley on the Olympic Peninsula, was once known as the rainiest spot in the United States, having recorded 184 inches of precipitation many years ago when a rain gauge existed there. The rain gauge is long gone, but the area still gets a lot of moisture, and the 184-inch figure for a 12-month period still shows up in some publications.

There's no rain gauge on the high slopes of Mt. Olympus, about 50 miles to the north of Oxbow, but that peak (elevation 7,976 feet) may be even wetter. It is believed that the mountain's windward slopes accumulate 200 or more inches of precipitation in a year's time.

Yet the town of Sequim, only 35 miles northeast of Mt. Olympus, lies in a "dry shadow" of the Olympic Range. Sequim receives less precipitation in a year's time than San Francisco, and irrigation of farm crops is a common practice in the Sequim area.

Ice and Snow

Besides its extremes of wet and dry weather, Washington has about 95 percent of all glacial ice in the 48 conterminous states below Canada. Most of the glaciers are on Mt. Rainier, Mt. Baker, Mt. St. Helens, Mt. Adams, Glacier Peak, and Mt. Olympus. Some of the world's most spectacular alpine scenery is found on those peaks and adjacent mountains.

Four major climatic zones are found on Mt. Rainier and other high peaks in the state. Those zones from low to high levels are: (1) Humid Transition, (2) Canadian, (3) Hudsonian, and (4) Arctic-Alpine.

The Humid Transition Zone extends from sea level to about 3,000 feet (sometimes beyond 3,500 feet in sheltered valleys) and contains the finest specimens of Douglas fir, western hemlock, western red cedar, spruce, and other trees prized for their timber. A variety of hardwoods also grow in this zone, including maples, alders, and dogwood.

The Canadian Zone, extending from the upper limit of the Humid Transition Zone to approximately 5,000 feet, contains some of the species found at lower elevations, as well as white pines, silver and noble firs, and Alaska yellow cedar.

The next zone, the Hudsonian, in the changing climatic picture is also below the timber line, but it doesn't have as many trees. The dominant trees are alpine fir, mountain hemlock, and whitebark pine. This zone, extending from about 5,000 to 6,500 feet, does have the most spectacular displays of flowers found in Washington's mountains. A burst of floral splendor is seen from late July through August, and is unsurpassed for beauty anywhere in the world.

The highest zone, the Arctic-Alpine, has the harshest weather—fierce blizzards, whiteouts, and other hazards for

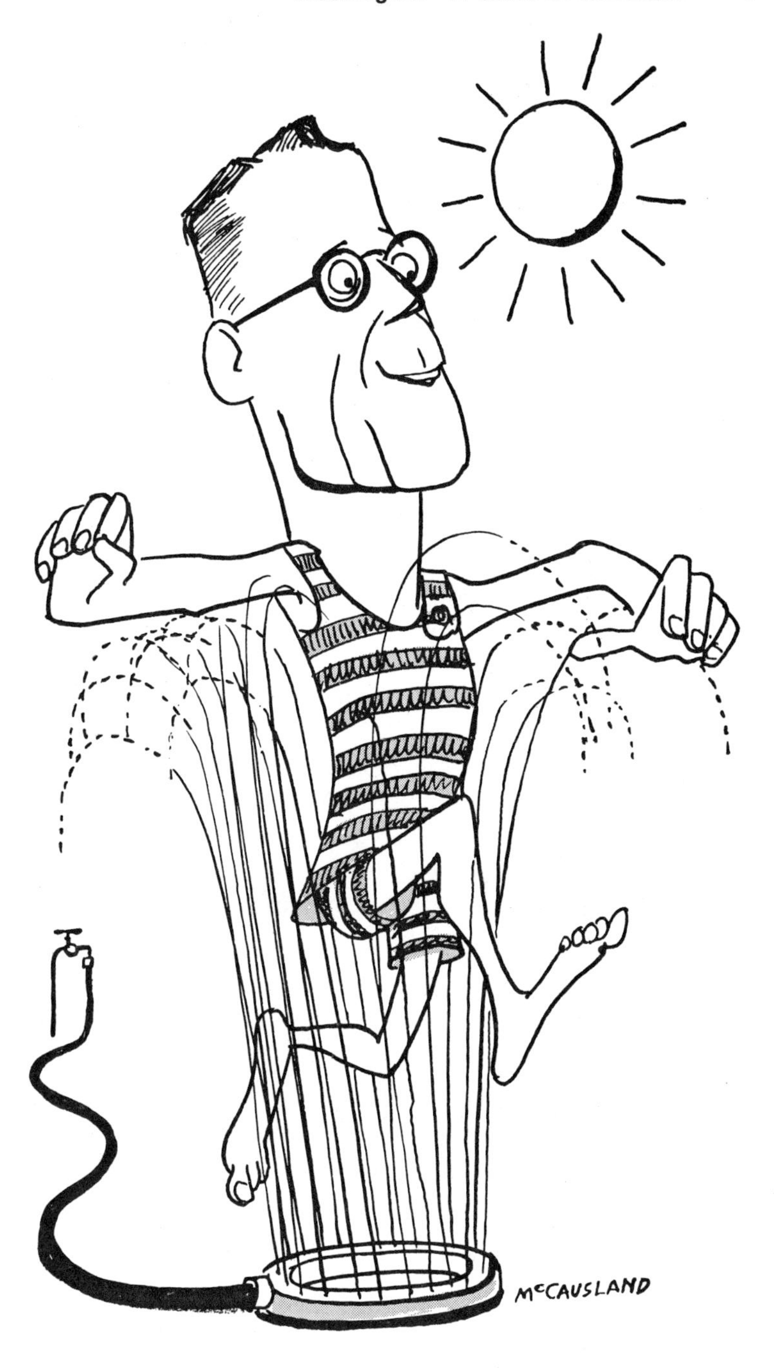
McCAUSLAND

climbers—and takes in all the high areas above the tree line. Most of this zone features perpetual snow and ice and barren rock. However, a few hardy floral species have been seen up to about 10,000 feet.

The average elevation of the Cascade Mountains is approximately 7,200 feet, although a number of peaks are considerably higher. The four climatic zones in the mountains are, therefore, most pronounced on Mt. Rainier, which rises 14,410 feet above sea level.

Mt. Rainier is situated some distance west of the main Cascades crestline, hence its south and southwest flanks catch the full fury of storms moving northward in Western Washington. Paradise Ranger Station in Paradise Valley (elevation 5,550 feet), holds the American record for the greatest total snowfall in a year's time. The greatest cumulative depth of snow was 1,122 inches (93½ feet) during the snow season of 1971-72. Other totals in excess of 1,000 inches measured at the same station were 1,000.3 inches, 1955-56, and 1,027 inches in 1970-71.

Rainier's heavy snows usually commence in October and continue through April.

Snowfall totals at Paradise Valley during the big snow year of 1971-72 were: July, 1971, 3 inches; August, none; September, 9 inches; October, 58.5 inches; November, 122.5 inches; December, 227 inches; January, 1972, 277 inches; February, 159 inches; March, 109.5 inches; April, 143.5 inches; May, 8 inches, and June, 5 inches. The snowfall year runs from July through June of the following year, inclusive.

Snowfall on the state's Olympic Mountains (the Coast Range) is peculiar in one respect. The snowline—the elevation above which there is snow the year around—lies at 6,000 feet on the western side of the range. That snowline is the lowest to be found in the 48 conterminous states.

The snowline on the east side of the Olympics, a drier region, is somewhat higher.

The Moderate West

Olympia, Tacoma, Seattle, Everett, and other Puget Sound communities lie in a favorable climatic region which has the state's greatest concentration of people and industries. The Olympic Mountains to the west afford considerable protection from some furious storms that make landfall on the Washington coast. And the Cascade Range effectively bars many inland cold waves from spilling into the Puget Sound region. The Cascades also hold back considerable hot air that might otherwise seep into Western Washington during the summer.

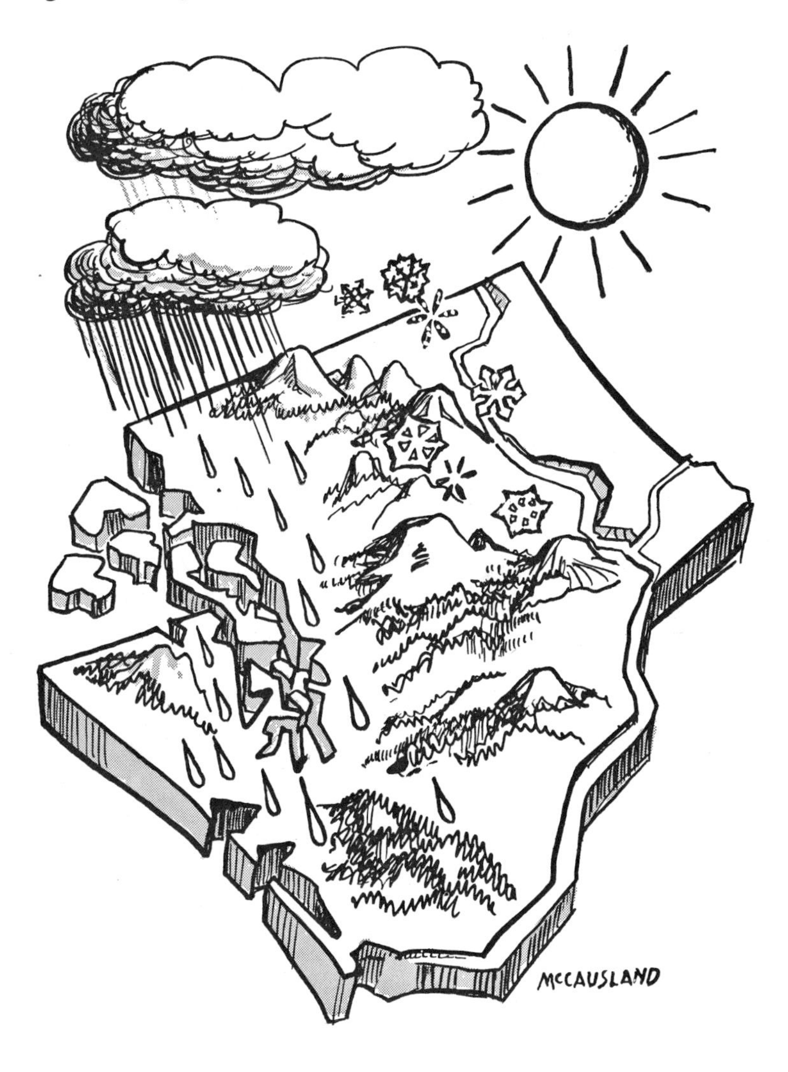

Extremes in the East

The weather east of the Cascades is radically different from that west of the range. There's a lot of desert in Eastern Washington, although some of that desert has been transformed into productive farms, orchards, and cattle country through irrigation.

Eastern Washington comprises all of that portion of the state lying east of the crestline of the Cascade Mountains, an area of 44,600 square miles, or five-eighths of the state. It includes the Great Basin of the Columbia River.

The Columbia River originates in Canada and flows through narrow valleys and steep-walled canyons from the northeast to the southwest in Eastern Washington, then turns to the west-southwest to form part of the border between Washington and Oregon. The river veers again, first northwest and then west-northwest before emptying into the Pacific Ocean.

The region east of the Columbia River is largely a rolling plateau ranging in elevation from 700 to 2,700 feet above sea level. The Snake River, originating in the Rocky Mountains, and numerous tributaries of the Snake, contribute to the large volume of water that eventually reaches the Pacific Ocean via the Columbia River.

The Palouse River, with its tributaries, and a portion of the Snake River (into which the Palouse empties) drain a vast region in Southeastern Washington known as "the Palouse Country," one of the nation's most productive wheat-growing areas. The Palouse Indians were the original residents there.

Rainfall is usually sufficient in the Palouse Country and in the Spokane, Colville, and Pend Oreille Valleys, as well as near the Blue Mountains (a chain of peaks extending into Southeast Washington from Oregon) to grow crops without irrigation.

In areas nearer the Columbia River, in the Yakima and Wenatchee Valleys, and other valleys northward to the Okanogan region, rainfall is scant and irrigation is necessary for the successful cultivation of crops. These are fruit-raising centers, famed for apples, peaches, pears, and other tree crops.

Most of the air moving into Washington from the Pacific Ocean flows up and over the Cascade Range. It cools as it rises into the colder areas of the high peaks. That cooling causes most of the moisture to drop as rain or snow because the cool air can't hold as much moisture as warm air.

Another change takes place as the air flows down the east slopes of the Cascades. It warms itself dynamically by compression, thus increasing its capacity to hold water vapor. Consequently, there is a preponderance of dry air with scant precipitation over all the valleys of central Washington, but especially in the lower parts of the Douglas, Chelan, Grant, Franklin, Yakima, Klickitat, Benton, and Walla Walla Counties.

The mean annual precipitation in those areas ranges from 6 or 7 inches in the driest parts to 11 or 12 inches in adjacent regions.

This section of the state is subject to greater extremes of temperature than Western Washington. There is considerable sunshine except in the winter, and the air is dry most of the year.

From Ellensburg, at an elevation of 1,510 feet on the east side of the Cascade Range, to the summit of Snoqualmie Pass, elevation 3,010 feet, there is an increase of precipitation from approximately 9 inches yearly to slightly more than 91, in an airline distance of only about 50 miles.

East of the Columbia River there is a gradual increase of precipitation to more than 20 inches per year. Walla Walla, in the southeast corner of Washington, gets an average of 16.01 inches of precipitation in a year, and Spokane, near the Idaho border and approximately half-way between the state's northern and southern borders, averages 17.42 inches a year.

In contrast, Quillayute Airport, situated on the coastal plain between the Pacific Ocean and the Olympic Mountains (three miles inland and only 180 feet above sea level), gets drenched with an average of 104.99 inches of rain per year.

Washington Weather Records

Washington's highest temperature was 118 degrees

Fahrenheit, first recorded on July 24, 1928, at Wahluke in the vicinity of the Hanford Works, east of Yakima. That high was equaled on August 5, 1961, at Ice Harbor Dam, Walla Walla County.

The all-time low for the state was 48 degrees below zero on December 30, 1968, at Mazama and Winthrop, both in Okanogan County. Another interesting low reading was -43 degrees on December 29, 1968, at Chesaw, Okanogan County, near the Canadian border.

The state's previous all-time low, which had stood for over three decades, was -42 at Deer Park, near Spokane, on January 20, 1937.

(The record high for the United States was 134 degrees Fahrenheit at the Greenland Ranch, Death Valley, California, set on July 10, 1913. The all-time low for the nation was 80 degrees below zero at Prospect Creek, Alaska, on January 23, 1971.)

The greatest snowfall in Washington during one calendar month was 363 inches in January, 1925, at the Paradise Ranger Station on Mt. Rainier.

The least precipitation recorded anywhere in the state in a year's time was 2.61 inches at Wahluke in 1930.

But for just plain wetness in the form of rain, the championship for one calendar month went to Peterson's Ranch near Cougar in the Lewis River Valley of Cowlitz County in Southwest Washington. The figure was 57.04 inches, measured in December, 1933.

For both wetness and dryness, Washington can hold its own with any section of the nation. And although other sections of the country can claim both higher and lower temperatures, for sheer diversity of climate, you can't beat the Pacific Northwest.

Northwest Weather—Something for Everyone

DOWNTOWN Seattle is a heat island, warmer than the surrounding countryside. All other cities and towns of any consequence have heat islands, too. The extra heat, most pronounced at night, comes from sewerage lines, steam pipes, and other warmth-producing conduits underground, as well as from pavement, buildings, and automobiles. Furnaces all over town add their BTUs to the city's man-made urban climate. Like the average big city, our downtown temperatures are from one to three degrees higher than those in rural sections.

There is little difference in the summer between daytime temperatures in the city and in the country. The daily maximums average out for Seattle and Seattle-Tacoma International Airport during June, July, and August: June—70.2 degrees Fahrenheit downtown and 69 degrees at the airport; July—75.1 degrees downtown and 75.1 at Sea-Tac Airport (no difference); August—74.2 degrees downtown and 73.8 degrees at the airport.

The relative warmth of the city is most apparent at night in the summer because the large expanses of concrete hold the heat of the day long after temperatures have begun

McCAUSLAND

to slide in the suburbs. Maximum daily temperatures in the city, particularly in summer, are sometimes reached an hour or two after maximums have been recorded at outlying stations.

Seattle has had a high of 100 degrees twice, first on July 14, 1941, and again on June 9, 1955. Both readings were taken from the downtown office of the National Weather Service. That office was moved to the Lake Union Building in 1972.

Winter minimums in the downtown heat island and in the outlying areas show a greater contrast: December—38.9 degrees downtown and 35.5 degrees at Sea-Tac Airport, which is typical of many suburban sections; January—36.2 degrees downtown and 33.0 degrees at the airport; February—38.1 degrees downtown and 36.0 degrees at Sea-Tac. (Downtown figures are from records kept prior to the move to Lake Union.)

Elliott Bay and Lake Washington have a tempering effect on our winter temperatures downtown. The all-time low for the downtown office was three degrees above zero on January 31, 1893.

In many parts of the United States, cities get from five to ten percent more clouds and rain than the surrounding countryside, and lower humidities. Seattle is an exception, insofar as rainfall and humidity are concerned, because of the topography of Washington and the presence of large expanses of water near us.

Rainfall throughout the Puget Sound trough increases as one travels southward (with some notable exceptions due to the effect of the mountains). Seattle's downtown rainfall, for example, probably averages around 37 inches, compared to 38.79 inches for Seattle-Tacoma International Airport. (Again, the figure for downtown is not exact, since official readings are no longer taken in the heart of the city.)

Even within the city, rainfall varies. The Maple Leaf area gets 32.20 inches of rain yearly, Sand Point 33.15, the University of Washington 34.41, and Boeing Field 35.63.

Out in the Green River Valley the precipitation totals 36.50 inches per year. Tacoma gets 35.20 inches, and Puyallup 38.93.

Index, at an elevation of 532 feet in the foothills, can expect 85 inches of rain a year. Landsburg, at approximately the same level, gets 55.48 inches, and Lester at 1,626 feet is in the 60-inch category.

Cedar Lake (elevation 1,560 feet) is drenched with 102 inches of rainfall annually. Paradise Lodge, Mt. Rainier, gets 99.74 inches, and Mt. Baker Lodge chalks up 111 inches.

Many of the rain clouds are wrung fairly dry by the time they reach the east side of the Cascades. As a result, Ellensburg gets only 8.36 inches of rain per year, and Sunnyside usually has less than seven inches.

The Dry Shadow, Dry? Dry!

Two of the strangest oddities in the nation's climate are found less than 70 miles from Seattle.

The community of Sequim, between Port Angeles and Port Townsend, lies close to the center of a sunshine belt where the annual precipitation averages only 16.81 inches.

Yet less than 50 miles away, on a bend of the Wynoochee River, the rainfall exceeds 150 inches per year. And it's likely that some parts of the nearby Olympic Mountains get close to 200 inches of precipitation in a 12-month span.

Port Townsend, situated on the western tip of Quimper Peninsula, commanding the entrance to Puget Sound, receives a bit more rain than Sequim—a little over 18.34 inches annually—but is slightly warmer. The added warmth, just a fraction over one degree (annual average), isn't of major importance in the climatological picture, but proud Port Townsendites have been using it as "ammunition" in a friendly feud with Sequim.

Just 17 miles west of Sequim is the bustling city of Port Angeles, which gets around eight inches more precipitation per year than Sequim (on the average), yet approximately ten inches LESS than Seattle, where the annual mean is 38.79 inches.

The topography of the Olympic Peninsula accounts for a sunshine belt so close to one of the rainiest parts of the United States. The Sequim-Port Townsend area is nestled in the "dry shadow" of the Olympic Mountains.

Most of the storms that move into Washington from

the ocean have a southwest to northeast flow. The maritime air that hits the Olympics moves upward and cools. The higher it goes, the cooler it gets, and this orographic flow generates rainfall. The rains occur because the cooler an air mass gets, the less moisture it can hold, and the excess moisture falls as rain.

Oak Harbor-Coupeville area on Whidbey Island, directly across Admiralty Inlet from Port Townsend, shares the climate of Port Townsend.

The Oak Harbor-Coupeville precipitation averages less than 18 inches annually, which is only a couple of inches more than the annual total for Los Angeles. A number of other sections of Whidbey are equally as dry.

Snowfall in the Coupeville-Oak Harbor region totals only six inches per year, on the average. The deepest snows in the past 30 years fell in January, 1950, when the month's snowfall amounted to 27.8 inches.

During the summer and fall, low clouds or fog often form in the morning, over Whidbey Island, but the sun usually breaks through around noon.

Here are the statistics on Whidbey's cloudiness: totally clear days, 43 per year, on the average; partly cloudy days, 67 (that means some sun during each of those days); cloudy days, 255.

In addition to being dry, Whidbey Island is generally calm and temperate. In September, for example, a calm prevails over Whidbey Island 24 percent of the time, and the breezes are gentle (3 to 8 miles per hour) 49 percent of the time. Stronger winds may be expected in September as follows: 9 to 14 m.p.h., 17 percent; 15 to 23 m.p.h., 8 percent; and 24 to 35 m.p.h., 2 percent.

October's Whidbey winds average out like this: calm, 20 percent of the time; 3 to 8 m.p.h., 42 percent; 9 to 14 m.p.h., 19 percent; 15 to 23 m.p.h., 14 percent; and 24 to 35 m.p.h., 5 percent.

Even in November, the air is calm 15 percent of the time over and near the island, and the most prevalent winds are in the 3-to-8 m.p.h. range.

December and January are the stormiest months on Whidbey, just as they are in and near Seattle, but winds of

McCAUSLAND

gale velocity are the exception, not the rule, even in winter.

The highest temperature ever recorded in the Coupeville-Oak Harbor area over a 30-year period was 98 degrees on August 8, 1960. The lowest reading in that same period was three degrees on January 3, 1950. The low at Coupeville during the Big Freeze of November, 1955, which caused so much damage throughout Washington, was eight degrees on the 15th.

July and August rank exactly the same for warmth on Whidbey Island. The temperature averages 60.9 degrees during those months. The average maximum temperature during that period is 71.3 degrees in July and 71.4 in August.

Even Victoria, on Vancouver Island, gets partial protection from the north flank of the Olympics. That city's mean precipitation per year is 26.18 inches, which is 1.57 inches more than that of Port Angeles, but nearly eight less than Seattle's.

Now hear this: Henderson Lake on Vancouver Island's west coast, where Pacific storms sweep in with even more fury than in Washington, averages 262 inches of rain per year and has had over 323 inches in a year's time.

Tatoosh, which lies just off the tip of Cape Flattery, gets drenched with more than 77 inches of precipitation annually.

The Washington Coast—From Tatoosh to North Head

The stormiest section of the Washington coast is Tatoosh Island, just off Cape Flattery. A brief look at Tatoosh's history perhaps provides the best overview of that rough rock's weather:

"Gale began blowing at 6:45 a.m. and reached its maximum velocity of 93 miles an hour at 3:25 p.m., exceeding all previous records at this station."

That entry is part of a weather report for November 26, 1920, in the National Weather Service's log on Tatoosh Island. (Tatoosh Island is no longer an observation site; the Weather Service abandoned the island and established a station at Quillayute Airport, ten miles west of the town of

Forks on the mainland.) The writer added: "Some windows were broken at the lighthouse. The top of a water tank at the radio station was blown off and hurled against the side of the main building. The windmill, instrument shelter support, and stove pipe at the Weather Service station were wrecked."

That's not an isolated instance of rough going on the "Big Rock" whose 17 acres rise out of the ocean at the entrance to the Strait of Juan de Fuca. The Weather Service and Coast Guard men who were on duty there in recent times had more durable buildings and better equipment than the personnel of earlier days, but the island is still subject to the eternal attrition of sea and wind.

If roughing it off Cape Flattery is not your forte, pack your picnic anyway and head for one of Washington's other ocean beaches farther south. They are delightful—unexcelled climatologically in the spring, summer and early fall—and one of the most attractive parts of our coastal strip.

Seeing is believing, and most of us have been there, but for our visitors and all others who like buttressing facts, here is a report on the climate of Washington's ocean beaches:

In midsummer the average afternoon temperature on the coast ranges from 65 to 68 degrees. The nighttime temperature near the ocean is about 50 in the summer.

The ocean current off the coast of Washington reverses direction between winter and summer. In winter the Davidson Inshore Current moves north, and in summer the so-called California Current travels south.

The temperature of the water along the coast in our latitude ranges from 48 degrees in February and March to 58 degrees in August.

During July and August it is not unusual for weeks to pass with only a few light rain showers. The wind ranges from a gentle to a fresh ocean breeze. Offshore seas are eight feet high or higher less than five percent of the time in midsummer.

A clockwise circulation of air occurs in summer around a high-pressure cell in the atmosphere over the eastern Pacific Ocean. The windflow pattern results in northwesterly breezes for our coast.

The summer ocean air has a temperature of 55 to 60 degrees. It becomes warmer and drier as it moves inland.

A few miles inland afternoon temperatures in summer are in the 70s most of the time. On rare occasions hot, dry winds from Eastern Washington sweep westward, resulting in 90-degree weather along the coast. But the heat waves don't last long, and they are accompanied, fortunately, by extremely low humidities (30 percent or less).

In the latter half of the summer and fall, fog banks frequently form offshore, moving inland at night. The fog usually evaporates by noon of the following day.

Following are some of the highlights of Washington's coastal weather:

North Head, southwest tip of the state: record high, 97 degrees in July, 1906, and June, 1925; record low, 11, in January, 1888; fastest wind 113 miles per hour on January 29, 1921; average annual temperature, 50.3 degrees; average annual precipitation (most of which falls in winter), 58 inches; average number of days with heavy fog, 44; average number of thunderstorms per year, 3.

(The wind velocity is based upon the speed of the fastest mile of wind to pass a National Weather Service anemometer. Wind gusts during the Great Storm of January 29, 1921, were estimated at 150 m.p.h. in the North Head area.)

Grayland, southwest corner of Grays Harbor County: record high, 96 in July, 1961; record low, 11 in January, 1957; average annual temperature, 50.2; average annual precipitation, 76 inches.

Aberdeen-Hoquiam area: record high, 105 degrees in August (old data, year not given); record low, 6 in January, 1950; average annual temperature, 50.3 degrees, average annual precipitation, 84.54 inches. (Aberdeen's record high in recent years was 104 in July, 1942.)

West Coast weather throughout the Northern Hemisphere is milder than East Coast weather in comparable latitudes.

Mt. Rainier—An Arctic Isle

Traveling from the ocean beaches up Mt. Rainier's

slope is like going from the seaside to the Arctic because Mt. Rainier is an arctic island in a temperate sea. It is a glacial octopus. It is paradise in good weather and a bit of hell in bad. Its climates, soils, vegetation, and wildlife are so diverse that a lifetime of study would barely scratch the surface in probing this dormant fire-peak.

Changes in the flora of Rainier are, broadly speaking, similar to those one would observe in traveling from Puget Sound to the Arctic. There are four life zones within Rainier National Park—the Humid Transition, Canadian, Hudsonian and Arctic-Alpine. Each of the lower zones has its distinctive plants. Mammals and birds are less inclined to remain within a specific climatic belt, but even they have preferences.

The peak isn't, as often supposed, a part of the Cascade Range proper. Its summit is about 12 miles west of the Cascade summit crestline, and therefore within the Pacific slope's drainage system.

At one time (in the long, long ago) Rainier was higher than 16,000 feet, judging by the steep inclination of the lava and cinder layers on its flanks. The mountain in that prehistoric time was one of the most active volcanoes in the Pacific's "rim of fire."

Then a great explosion occurred, decapitating the peak and reducing its elevation by some 2,000 feet. The new summit consisted of a hollow crater (now snow-filled), surrounded by a jagged rim. There have been slight eruptions since—one in 1843, another in 1854, one more in 1857, and the last in 1870. Jets of steam still spurt from the mountain's summit, and there are hot springs on its lower flanks.

Volcanic ash is abundant on the upper slopes and sometimes it swirls in the wind, leading people to believe that Rainier is "coming alive" again. The swirling ash looks like smoke.

But don't be alarmed. The peak is dormant. Its inner fires are quiet.

Rainier's elevation was set at 14,408 feet in 1913 by a United States Geological Survey team. A resurvey in 1956, using more refined methods, disclosed an elevation of 14,410 feet.

The Humid Transition Zone on Rainier has plant life like that at sea level around Puget Sound. It extends to about 3,000 feet, featuring Douglas firs, western hemlock, lacy red cedars, bigleaf maples, Pacific dogwood, and numerous other trees, as well as Oregon grape, salal, flowering currant, lilies-of-the-valley, and a multitude of ferns.

On the west side, in sheltered valleys, this Humid-Transition Zone sometimes creeps past 3,500 feet.

Next is the Canadian Zone, whose forests are not so dense as those below. Typical trees are the Pacific silver fir, Alaska yellow cedar, noble fir, and white pine. Among the plants are the queencup beadlily, bunchberry, pipsissewa, Oregon wintergreen, fool's huckleberry, and azalea.

The third zone, and perhaps the most interesting, is the Hudsonian, ranging from about 5,000 to 6,500 feet. This zone is covered with snow for most of the year (usually from November 1 to July 1). In consequence, the growing season is short and intense. Some of our favorite flowers, including the avalanche lily, grow in the Hudsonian climate.

Mt. Rainier's most spectacular wildflowers grow in the Hudsonian meadows, which completely circle the peak. The flowers seem to "explode" all at once into their colorful display, thanks to the sun's warmth in July and the abundant moisture from melting snowbanks.

Many of the trees and flowers in this zone are the same as, or similar to, those found in northern Canada and Alaska.

Typical Hudsonian trees on Rainier are alpine fir, mountain hemlock, and whitebark pine. The next time you look at any of the Hudsonian trees on the mountain (at Paradise Valley, for example) note how they grow. The limbs are short, making it impossible for tons of snow to cling in the branches and possibly tear them from the trunk.

The Arctic-Alpine Zone includes all areas above the upper limits of tree growth. The altitudinal range is from about 6,500 feet to the summit. Only the lower portions have plant life, as the greater part of the zone is characterized by barren rock, perpetual snows or ice. The majority of flowering plants are in the belt from 6,500 to 7,500 feet, although a few species are found up to 10,000 feet.

Even at the crater's rim—about 14,300 feet—a few mosses and lichens are found where steam vapors give a little warmth.

Distinct lines cannot be drawn between the zones. The changes from one to another are gradual, with intermingling of zonal life.

The heaviest snow on Rainier falls between the 5,000 and 10,000-foot levels because most of the clouds lie within that belt. The mountain's summit is often in the clear while blizzards are raging farther down the peak.

More than 130 species of birds and 50 species of mammals have been seen in Rainier National Park. Raccoons, chipmunks, marmots, mountain goats, deer, and bear are common.

The Oregon jay (also known as camp robber) and Clark's nutcracker are among Rainier's best-known birds. The nutcracker is built like a crow but in coloring and habits it resembles the jay. The nutcracker is numerous at both Paradise Valley and Sunrise Park and is the tamest bird at those parks, often accepting food from outstretched hands.

Washington's Meandering Snowline and Warm (Well, Sort of . . .) Glaciers

Little-by-little, as the year wastes away, the snowline drops in Washington's mountains. Early in September one must look closely to see the powdery white stuff lying in the shadows on peaks above 6,000 feet in the Cascade Range. But it doesn't take long to bring the snowy blanket down to the ski slopes.

The perpetual snowline in Washington's Cascades lies between 7,000 and 8,000 feet. The snows are eternal above that mark.

The Olympic Mountains, our coastal range, hold the distinction of possessing the lowest snowline in the United States, south of Alaska. Winter's perpetual threshold there lies between 5,500 and 6,000 feet as the result of a cloudy, moist climate.

The Olympics are the first high-rise barrier on our continent in this latitude, and they catch the brunt of most ocean storms. Precipitation is generous in those mountains, occurring most of the time as rain in the lowlands and snow at high elevation.

In the mountains of southeast Alaska, one of the wettest areas in the West, the permanent snowline is 4,500 to 5,000 feet above sea level. Other mountains, directly east in British Columbia, have a snowline about 2,000 feet higher.

Antarctica is the only region on earth where the snowline is at sea level. Contrary to widespread belief, snowfall isn't heavy in the Antarctic. The edge of the Ross Ice Shelf at 78 degrees South Latitude gets about one foot of snow per year and the South Pole about six inches.

In contrast, Stampede Pass in Washington can expect over 400 inches of snow in a year, and Paradise Valley, Mt. Rainier, around 582 inches. The elevation at Stampede Pass is 3,958 feet, and that at Paradise, 5,550 feet.

Even at Darrington in the Cascade's foothills (elevation 550 feet), the snowfall averages more than 50 inches per year.

The Arctic has a snowline at sea level in the winter only. In summer considerable melting of both snow and ice may occur. A few years ago scientists on an ice island in the Arctic Ocean reported that temperatures rose above freezing in mid-May and by mid-August all snow, plus a foot or more of ice, had melted.

Washington's snowline usually drops to a line somewhere between 1,500 and 2,000 feet in winter, and in extremely cold winters it plunges considerably below 1,000 feet. This is a reference to the downslope edge of a long-lasting snow blanket and not to the occasional snow cover we experience here at sea level.

Topography and exposure to the weather have a lot to do with the meanderings of the snowline.

Greenwater, Washington, a community alongside Highway 410, 18 air miles east-southeast of Enumclaw, is typical of many areas just above winter's snowline.

Greenwater's average yearly snowfall is 77 inches. Snow can be expected from the latter part of December through March, and sometimes into April.

In 1955-56, a cold, snowy winter, Greenwater had 17 inches of snow on the ground by November 21. The depth of snow on the ground had reached 46 inches by February 17, the greatest amount on the ground at any one time. But the winter's total snowfall reached 154 inches.

The snowline rises nearly 8,000 feet in the 450 miles between Mt. Olympus, central peak in the Olympic Range, and Mt. Shasta, California. And it continues to rise as one proceeds toward the equator.

In central Mexico, land of high mountains, the snowline lies between 16,000 and 17,000 feet. The highest snowline, around 21,000 feet, is in the Andes of South America, more than 1,200 miles south of the equator. That extremely high snowline is in the dry Horse Latitudes. When the snows of winter don't totally melt away, glaciers, like those on Mt. Rainier, form and survive the seasons' changes. But believe it or not, some glaciers are "warm," while others are "cold." Ours in Washington, totaling about 135 square miles of ice, are on the warmish side. Seems strange, doesn't it? Those are scientific classifications of course, and warm means warm only in relation to a chillier chill. A "warm" glacier is one that heats up to the melting point, 32 degrees Fahrenheit, at some time during the summer. A cold glacier, on the other hand, maintains a temperature below freezing throughout most of its mass most of the time.

In some parts of the world, notably the interior of Greenland and the heart of Antarctica, the ice is far below freezing. Some cold glaciers may have a limited amount of the surface ice melt under the sun's influence, but warming is never sufficient to bring the entire mass to the melting point.

The outer edges of Antarctica's and Greenland's ice are heated at times by air moving in from warmer parts of the earth, but as the air travels inland, the ice cools it. The central portions of those monstrous ice-cakes are therefore exceedingly cold.

The world's burden of ice is believed to total somewhere around nine million cubic miles, covering ten percent of the land. Antarctica has from seven million to eight million cubic miles of ice (4,900,000 square miles), and Greenland has about one million cubic miles of frozen water (670,000 square miles).

Most of the ice in the United States is in Alaska, where the total approximates 20,000 square miles. Eighty percent of the glaciers in the other 49 states are in Washington. Of Washington's 135 square miles of ice, nearly half is on five mountains—Rainier, Baker, Glacier, Adams and St. Helens. The rest is in the northern portion of the Cascade Range, exclusive of Mt. Baker and Glacier Peak.

The remainder of America's glacial age is in Wyoming, seventeen square miles; Oregon, eight square miles; California, six square miles; Montana, five square miles, and Nevada, Colorado, Idaho and Utah, whose totals are small.

Our Northern Cascades, mainly from Glacier Peak northward, are the most rugged mountains in the Pacific Northwest, featuring sharp ridges, deep cirques and hanging glaciers. These mountains contain much granite, gneiss, and other metamorphic rock in contrast to the Southern Cascades, which have heavy layers of dark lava.

Metamorphic rocks have undergone considerable change due to pressure, heat, or the action of water in any of its forms, resulting in a compact, more highly crystalline structure. Gneiss (pronounced like nice) is a laminated or foliated metamorphic rock corresponding in composition to granite and certain other rocks.

The glaciers in Washington are not remnants of the ice age, having come into being separately. Glaciers like those on Mt. Rainier form and remain in existence when the snows of winter aren't totally wiped out by summer's warmth. Our Cascades are one of the snowiest mountain ranges on earth.

Once created, a glacier will survive if its source is above the snowline and the snows continue to fall. Two of the world's youngest glaciers were "born" during this century in a crater on Mt. Katmai, Alaska. Katmai erupted furiously in June, 1912. The explosion expanded a crater at the summit and severed the top sections of several old glaciers—huge ones—which originated on or near the mountaintop.

But parts of the crater's new surface were still above the snowline. The snows piled up year after year, finally compacting into ice masses on two separate shelves formed by rockslides.

British Columbia, Canada to the North

Two of Canada's principal climatic zones, the Pacific and the Cordillera, are found in British Columbia, and a third, the Northern, takes in the extreme northeast corner of the province.

The Pacific Zone

Vancouver and Victoria, the two biggest cities in British Columbia, are situated in the Pacific Zone where winters are mild and summers moderate. Both of these metropolitan areas are sheltered by the Coast Mountains from most arctic cold waves that sweep southward into regions farther east.

The Cordillera (referring to mountains) encompasses most of the mountain systems in the province. This zone extends from the Coast Range, an extension of the Cascade Mountains in Washington, to the eastern flanks of the Canadian Rockies and offers many contrasts in weather.

At its northern boundary with the Yukon Territory, British Columbia spans nearly 500 miles of mountains, plateaus and plains, plus a small portion of the Great Plains east of the Rockies. The Cordillera is by no means confined to British Columbia; it runs straight through the Yukon into Alaska.

The Pacific Climatic Zone takes in Vancouver Island, the Queen Charlotte Islands, and other insular portions of the province, as well as a narrow coastal belt on the mainland. The mainland belt extends eastward only a few miles in some places, and nowhere for more than 100 miles. Temperatures in that zone rarely drop below zero degrees Fahrenheit in winter and seldom rise above 90 degrees Fahrenheit in summer. However, precipitation is not equally distributed.

Like the mainland, Vancouver Island catches the full

McCAUSLAND

fury of many Pacific storms. Henderson Lake, at the end of an arm of Barkley Sound, on the island's west side, is the wettest spot among reporting weather stations in America, averaging 262 inches of precipitation per year. This average is for a span of 14 years.

But Victoria, the provincial capital, at the southern tip of the same land mass, lies in a protected area known to meteorologists as a "dry shadow." There are numerous dry shadows in British Columbia, created by nearby mountains whose windward slopes capture much of the rain and snow that might otherwise reach those dry areas. All dry shadows are on the lee side of mountains.

The mean annual rainfall at Victoria International Airport amounts to 31.93 inches, and the average snowfall in a year's time is less than 18 inches. The Victoria Gonzales Heights station gets even less precipitation: mean rainfall, 24.79 inches; snowfall, 13.93 inches.

The mean temperature for the year at Gonzales Heights is 49.6 degrees Fahrenheit, and the averages for February, the coldest month, and July, the warmest, are, respectively, 41.2 and 59 degrees.

Bright sunshine at Victoria in a 12-month span exceeds 2,000 hours, on the average.

Vancouver, British Columbia, Canada's leading seaport and industrial area in the west, is similarly blessed with moderate weather, although her skies are a bit grayer and the precipitation higher than at Victoria. The mean annual rainfall at Vancouver International Airport totals 40.07 inches and the snowfall slightly over 20 inches.

The mean yearly temperature at the Vancouver Airport is 49.5 degrees, not even a degree higher than the figure for Victoria's Airport, but more than three degrees higher than the mean for Gonzales Heights. The summer months at the Vancouver Airport station also are slightly warmer than Victoria's summers. Both cities rightfully boast of having balmy weather in the spring, summer, and fall—without high humidities to cause discomfort. Thunderstorms are infrequent.

The growing season is long at low elevations in most

parts of British Columbia's Pacific Zone, averaging 217 days at Vancouver and more than 275 days at Victoria.

Even Prince Rupert, more than 500 miles northwest of Vancouver (at 54 degrees 17 minutes North Latitude) has much milder weather, on the average, than inland regions of the province. The growing season there (counting from the last frost of spring to the first frost of fall) averages 197 days. However, Prince Rupert gets drenched with more than 90 inches of rain in an average year.

Central British Columbia

Vastly different conditions prevail in central British Columbia, a region of plateaus, high plains, and river valleys where elevations range from 2,000 to 4,000 feet. Prince George, a town in the great Fraser River Valley within that expansive area, reports the following data: mean precipitation per year, 15.51 inches; snowfall, 66.5 inches; average date of springtime's last frost, June 17; average length of the growing season, 68 days, and mean number of days with freezing temperatures, 202.

Victoria's extremes of temperature over a long span of years were: 95 degrees Fahrenheit in summer and two degrees below zero in winter; Vancouver's—92 degrees high, two degrees low; Prince Rupert's—88 degrees high, minus six degrees below zero low and Prince George's—102 degrees high and minus 58 degrees below low.

The coastline of British Columbia is indented with many fiords, some of which penetrate into the heart of the lofty Coast Range. A number of deep river valleys also cut through the mountains. The Fraser River, originating in the Rockies and emptying into the Strait of Georgia at Vancouver, is one of these huge rivers.

In some of the valleys which the rivers have gouged through the mountains in order to reach the ocean, there is a transition from maritime climate at the mouths of the rivers to continental climate as you move back toward the source. The continental weather is warmer in summer, colder in winter, and drier.

The Coast Mountains, in combined height and area,

constitute the greatest mountain mass in all of Canada. Some of the peaks are over 10,000 feet in elevation, and one, Mt. Waddington (elevation 13,260 feet), is higher than any of the peaks in the Canadian Rockies. This range, therefore, performs splendid service for the coastal areas in turning back most of winter's bitterly cold continental air.

The western portion of British Columbia's southern interior has a number of deep valleys entrenched in a plateau whose average elevation is 4,000 feet. The eastern part of the southern interior sweeps to higher elevations in several mountain ranges, dips downward to the Columbia River Valley, then rises sharply once more to the western flanks of the Rockies.

The general north-south trend of the valleys in the western section of the southern interior often results in the movement of arctic air from the north during the winter. At times, too, the area is chilled by cold continental air that drifts westward out of Alberta through passes in the Rocky Mountains.

Numerous recreational lakes lie in that southern interior of the province, and many farms and orchards also exist there. The Okanagan Valley, in particular, is widely known as a fruit-growing region. Precipitation is scanty, but irrigation systems have been installed in some places, utilizing water from mountain streams. Winters are cold and summers warm with low humidity, making the climate ideal for fruit crops.

The central interior of British Columbia, embracing an area of about 87,000 square miles, includes the upper portions of the Fraser and North Thompson Valleys, as well as the Skeena River Valley. The Cariboo Mountains and the northern Monashee Mountains are on the east side of this area of big rivers, numerous lakes, deep valleys and rolling plateaus. Prince George and Quesnel are in the central region.

Here summers are short and cooler than those of the southern region, and precipitation is not heavy. Most weather reporting stations of the central interior have mean temperatures below freezing for five months or more. However, temperatures in the summer have occasionally exceeded 100 degrees. Quesnel's maximum was 105 degrees Fahrenheit.

In the upper Skeena River Valley, the growing season ranges from 50 to 70 days. In the upper Fraser River Basin the frost-free season ranges from 35 to 100 days.

Northern British Columbia

Northern British Columbia has long, cold winters and short summers. Precipitation is moderate. Outbreaks of polar ice have occurred at all seasons in that part of the province.

Some of British Columbia's mountains receive enormous amounts of snow. A weather station at Kildala Pass in the Coast Mountains reported an annual seasonal snowfall of 809 inches and once had 880 inches in a year's time.

In the valley bottoms of the province's southern interior, the winter snowfall usually ranges from 30 to 50 inches, and the amount increases to between 80 and 150 inches on the west slopes and tops of the uplands.

At Glacier, 4,094 feet above sea level in the north Columbia Basin, snowfall averages 342 inches during the snow season, and at heights above 6,000 feet—even in southern British Columbia—the snow cover lasts all year.

The highest temperature for British Columbia was 110 degrees Fahrenheit reported at two places: Greenwood, south of Penticton, and at a Skagit River station. Both points are near the border between Washington and British Columbia.

The all-time low was 61 degrees below zero, recorded at Vanderhoof in the central interior. Canada's greatest extremes of temperature were 115 degrees Fahrenheit at Gleichen, Alberta, on July 28, 1903, and 81 degrees below zero at Snag, Yukon Territory, on February 3, 1947.

The northeast corner of British Columbia, lying east of the Canadian Rockies, comprises the Fort Nelson and Liard River Basins in the north, separated from the Peace River country to the south by a plateau rising to 4,000 feet. The winters are long and cold, the summers cool and short, with moderate precipitation. Fort St. John in that part of the province has had temperatures as low as minus 53 degrees Fahrenheit.

British Columbia, like Washington State, is truly a land of many climates.

Oregon to the South

Oregon, lying mostly between 42 and 46 degrees North Latitude, has a variety of climates within its 96,981 square miles. Major influences on the weather are the Pacific Ocean, the Coast Range, and the Cascade Mountains, although other topographical factors also are involved.

Winter minimum and summer maximum temperatures are greatly moderated in Western Oregon—west of the Cascades by air masses moving inland from the ocean. The moist, mild air is a moderating influence east of the Cascades, too, but to a lesser degree.

The Coast Range, comparable in location and north-south orientation to Washington's Olympic Mountains, captures much of the precipitation falling out of rain clouds moving inland. The low-lying coastal strip is also watered abundantly by Mother Nature.

For example, the city of Astoria, near the mouth of the Columbia River, has a mean annual precipitation total of 66.34 inches, whereas Portland, 65 miles from the coast, averages 37.61 inches per year. The difference is due entirely to location. Astoria and the ocean are separated only by several miles of sand dunes, but Portland gets protection from the Coast Mountains. Portland is approximately midway between the Coast Range and the Cascade chain, each about 30 miles distant.

The weather of Western Oregon is similar in many respects to that of Western Washington, although there are some major differences. The coastal mountains in both states get considerable precipitation out of storms off the ocean, and the Cascade Mountains, loftier than the coast ranges, protect the western sections of both Oregon and Washington from most of the arctic air that filters into regions east of the Cascades.

There is, however, one channel through which cold air in winter and hot air in summer sometimes drifts from the east side of the Cascades to the Portland, Oregon, and Vancouver, Washington, areas. That escape route is the Columbia Gorge, a mighty cut through mighty mountains, created over eons of time by the Columbia River.

The eastside air, spilling through the gorge, usually is warmed in winter and cooled in summer by contact with moderate ocean air. But if the east-to-west push is intense, Portland and other communities close to the lower Columbia River get a taste of eastside weather.

Portland, two degrees farther south than Seattle, has had summer temperatures as high as 107 degrees Fahrenheit and a winter low of three degrees below zero. Seattle's all-time high was just 100 degrees, and the low, zero.

However, mean temperatures for the three winter months are almost the same for the two cities, and Portland's summers are warmer, so the big Oregon metropolis has a higher yearly average than Seattle—52.6 degrees Fahrenheit compared to 51.1 at Seattle-Tacoma International Airport.

Portland also gets slightly less precipitation, on the average, than Seattle. The annual means for the two cities, respectively, are 37.61 and 38.79 inches. Approximately 88 percent of Portland's precipitation occurs from October through May. The driest months are July and August.

Destructive storms are infrequent in the Portland area. Surface winds seldom reach gale force, and only twice have National Weather Service anemometers measured winds higher than 75 miles an hour. Thunderstorms also are infrequent, occurring only about once a month through the spring and summer months.

Heavy downpours are also uncommon. Almost all the rains are gentle, a fact widely publicized in the early days of Oregon Territory exploration. The gentleness of the precipitation gave rise to an expression still heard throughout Oregon and Washington, "It's only an Oregon mist."

The Portland area has a long growing season, sometimes lasting for more than 200 days, hence rural areas in that region are farmed extensively for berries, vegetables, and a variety of fruit. The seed industry is a big one, too.

One of the remarkable features of Western Oregon's

McCAUSLAND

topography is the Willamette Valley, a broad expanse of lush countryside between the Coast Range and the Cascades. The Willamette River, which flows through Portland into the Columbia, has numerous feeders of its own, the largest of which rise in the Cascade Mountains.

Salem, the state capital, is situated in the middle of the valley, about 60 miles east of the Pacific Ocean. The mean annual precipitation at nearby McNary Field is 28.01 inches, but precipitation is greater in the hills and mountains, resulting in excellent growing conditions for Douglas fir, hemlock, cedar, and other timber trees.

The valley's mild temperatures, long growing season, and fertile soil also are ideal for a great variety of farm and fruit crops. So agriculture, like lumbering, is big business in that region of Western Oregon.

Salem has had a high temperature of 108 degrees Fahrenheit and a low of 12 degrees below zero, but such extremes are unusual.

Eugene, home of the University of Oregon, is situated at the upper (southern) end of the Willamette Valley with mountains to the east and west and low hills on the south. Rainfall here averages 42.56 inches per year. The first fall rains usually come during the second or third week of September, with gradual increases until about the first week in January, after which there is a slow decrease. July and August are normally quite dry, sometimes passing without any precipitation.

Eugene's extremes of temperature were 106 degrees Fahrenheit and 12 degrees below zero, but these are exceptions to the rule. The normal weather is pleasantly mild. Perhaps you have heard sports announcers on radio or television tell you that Eugene is one of the West's most beautiful cities. They aren't exaggerating.

Medford, in Southern Oregon (at 42 degrees 22 minutes North Latitude), is in a mountain valley formed by the famous Rogue River and one of its tributaries, Bear Creek. The city has a climate of marked seasonal characteristics. Late fall, winter, and early spring are damp, cloudy, and cool under the influence of marine air. Considerable warming occurs in late spring and continues into early fall.

Medford has had temperatures of 100 degrees Fahrenheit at one time or another in May, June, July, August, and September. Medford extremes were 109 degrees Fahrenheit and 6 degrees below zero.

The Siskiyou Mountains separating Oregon and California put Medford in a so-called "rain shadow," resulting in light precipitation. The mean annual total is 20.64 inches, but the amount has been considerably less in dry years.

Eastern Oregon's climate is essentially the continental (inland) type, with cold winters, hot summers, and light precipitation. But unlike the humid areas east of the Continental Divide, the relative humidity is usually low. Thus, the air is invigorating and much less depressing than many regions in the Midwest and on the East Coast.

Oregon's mean annual temperature is highest, about 56 degrees Fahrenheit in the Snake River Canyon in the extreme northeast corner of the state and is above 54 degrees along the Columbia River in the north. These temperatures correspond with yearly means in Eastern Texas and Eastern Kansas.

There are some elevated districts in the central portion where the yearly mean is around 40 degrees Fahrenheit, corresponding to the climate of North Dakota.

Eastern Oregon's Columbia Basin (which also extends into a vast area of central Washington) is the state's principal wheat-growing region. Hood River County in the Basin is one of the West's principal fruit-growing regions, famed especially for apples, pears, and cherries. Fruits are also raised commercially in other counties adjacent to the Columbia River.

Umatilla County, in particular, produces many fruits, as well as vegetables and wheat. Beef cattle, sheep, and poultry also make substantial contributions to the economy of the Basin and bordering regions.

Wheat ranches are also extensive in Gilliam, Morrow, Sherman, Wasco, Wheeler and other counties; alfalfa is also a big crop.

Oregon's highest temperature was 119 degrees Fahrenheit, first recorded on July 29, 1898, at Prineville, Crook

County, and again on August 10, 1898, at Pendleton, Umatilla County. The lowest reading was 54 degrees below zero recorded twice in February, 1933—first at Ukiah, Umatilla County, on the 9th, and equaled the following day at Seneca in Grant County. All of these hot and cold spots are in Eastern Oregon.

Crater Lake, situated in Oregon's southern Cascade Mountains, has chalked up most of the state's snowfall records: the greatest total in 24 hours, 37 inches on January 27, 1937; the greatest amount of snow out of one storm, 95 inches January 15-19, 1951; the greatest total in one calendar month, 256 inches, in January, 1933; and the greatest amount in a year's time, 879 inches during the snow season of 1932-33.

The lake, 6,176 feet above sea level, occupies the huge crater left when Mt. Mazama blew its top off in a series of eruptions, perhaps only 8,000 years ago. Crater Lake National Park is one of the many attractions that make Oregon a scenic wonderland. And the state's numerous climates offer something for everyone.

3

Northwest Weather— Unique

THE rains come in the late fall, winter, and early spring because the moist air gets cooler as it blows inland. Important factors in the cooling of the air are the Olympic and Cascade Mountains. When the ocean air reaches the mountains, it is forced upward and becomes cooler. Cold air can't hold as much moisture as warm air, so it rains, provided the air mass is laden with water vapor.

Because the Olympic range is low, some of the ocean air moves eastward without dropping all of its moisture on the Olympic Peninsula. A lot of Pacific air also moves inland to the south and north of the coastal mountains.

Many storms make their landfalls in the low country at or near the mouth of the Columbia River, then move into this area through the Puget Sound trough.

The counterclockwise movement of air in low pressure systems is often a factor in this south-to-north movement of wind, clouds, and rain. The center of a "low" may be due west, or even northwest of us, over the ocean. The wind on the western side of the low will be moving from north to south, but eventually it turns and heads north in a great sweep.

McCAUSLAND

Many-a-time we find ourselves in the direct path of the bluster on the east side of those winter "lows." The Cascade Mountains help deflect some of the storminess toward us, but they capture a lot of the moisture for their own wintry mantles of snow.

The Olympic Mountains catch more rain (on their windward slopes) than the Cascades, except for pockets of exceptional wetness in the latter range. But the higher inland peaks like Mt. Rainier, Mt. Baker, and those along the spine of the Cascades accumulate more snow.

Seattle's winter warmth comes from rain and clouds. The formation of precipitation causes latent heat of condensation to be released in the atmosphere. The clouds keep heat from being radiated away from the earth and lost irretrievably in space. Much of the radiating heat hits the clouds and is bounced back to us.

Seattle's coldest weather occurs when the state is under the influence of continental polar air. This air is dry, dense and chilly. On such occasions there's usually a high pressure cell in this vicinity which acts like an immovable football line to ward off warmer Pacific air.

Most of the Arctic air moves into the United States on the east side of the Rockies. Occasionally Eastern Washington gets a blast of the chill air which has moved out of British Columbia through mountain valleys.

The Cascades protect us from almost all of the polar outbreaks, enabling maritime weather to prevail here. Maritime weather prevails here in the summer as well as in the winter.

If you live near Puget Sound, you have enjoyed gentle, refreshing breezes on warm days. You've probably noticed that the flow of air off the water is most pronounced in the afternoon when it's most appreciated.

This phenomenon is the sea breeze, caused by the heat. Its opposite is the land breeze which flows toward the water at night.

The daytime breeze off the water gets its start when air over the land becomes hot and begins to expand. The hot air rises in obedience to a law of physics, and cooler air over the Sound moves in at low levels to take its place.

It doesn't take long, though, for the cool air to get warm as it passes over hot earth and hotter masses of concrete, asphalt, brick, and steel in cities like Seattle, Bremerton, and Everett.

The result is a continual rising of warm air and its continual replacement, at least while the sun is shining. The land air spreads out broadly and some of it bends back toward the water where it originated.

This creates a circulatory pattern shaped somewhat like an egg, with the air on the underside moving toward land and the air overhead going to sea.

Sea breezes usually are less than 2,000 feet deep. They are strongest close to the shore, but may be felt several miles inland if enough push develops in the onshore flow.

At times Lake Washington produces a soothing summer breeze which develops in the same manner as a sea breeze, although on a smaller scale. Smaller lakes may have domes of cool air over them during a warm spell, but there isn't likely to be enough contrast between their air masses and the air over adjacent land to create a big circulatory whirl.

A land breeze, flowing toward water, is likely to occur at night because the earth cools much more quickly than the water after dark. The warmer night air over Puget Sound (or any other large body of water) begins to rise, just as the hot land air rises by day, and the denser air off the land rushes in to take its place.

Large-scale movements of air will, of course, prevent either sea or land breezes from developing. Our northerly breezes here on many summer days are crossing both land and water, headed for regions in Oregon and California where heat lows exist. Such flows of air are vast enough to wipe out purely local winds.

But sea breezes occur often enough here to be recognizable as local phenomena. They're welcome, too.

Trees, as well as the mountains and the sea, are climate-makers in the Northwest. And the cutting of trees will change the climate. But as long as we let nature renew the forests, we aren't going to alter the weather drastically. Cutover land, stripped down to bare soil, becomes a hotspot in summer. The earth dries and the humidity (perhaps already low) may drop to the point where even the ferns will shrivel and die. The same logged-off land may become colder by several degrees than nearby areas in winter.

But as the new trees come along—first the hardwoods such as maples and alders, then the pines, hemlocks or cedars—the forest floor takes on the sponginess of old. Dead ferns, dead leaves and rotting logs help hold back much of the precipitation that otherwise would erode the land as it rushed unchecked to the sea.

Thus we see the climate changing in little spurts, but overall remaining the same.

If it weren't for our evergreens in the mountains, we would have precious little water in reserve behind dams for drinking, lawn-sprinkling, irrigation of gardens and farms, and production of electricity during the summer and fall.

A study by German foresters indicates that trees help thrust air upward on the windward sides of mountain slopes, and thus increase rainfall over the forest by as much as six percent.

The trees also capture moisture from snow on their branches and from fog and rain, and thus contribute to their own well-being. And they do a good job of putting the brakes on the wind.

The mountains, the ocean, and even our trees all help to keep the Northwest climate mild, but there's a myth about the Japanese Current that needs exploding.

The current does NOT appreciably influence Seattle's weather. It is NOT responsible for mild winters in Western Washington, Western Oregon, and parts of British Columbia.

That statement will come as a shock to many residents of the Northwest, including some school teachers who have been telling generations of pupils about the current's warming effect.

The persistence of the myth about the current was illustrated when a climatological question was asked during a television quiz show for Northwest school children.

The question (for fifth-graders) was: "Why does Seattle have a much milder climate in winter than Augusta, Maine, even though Seattle is farther north?"

The answer listed as correct was: "The Japanese Current keeps the Seattle area warmer, while northern winds blow from Canada to Maine."

The truth is that the current is just a ribbon of slightly warm water in a cold ocean. The effect of the Japanese Current alone on the great air masses flowing over thousands of miles of ocean would be negligible.

At or near its source on the other side of the Pacific, the Japanese Current does influence the climate, but after it has pushed into the Far North, then circled around toward the United States it is a feeble force.

In fact, there isn't any distinctive Japanese Current off our West Coast. The true Japanese Current breaks up into a scattered system of water movements far out in the Pacific. The warm current off Western America is the California Current.

In general, the so-called warm currents of the eastern Pacific can't even be credited with a minor weather assist, one way or another. Everyone who has dipped a hand into the Pacific Ocean off our coast can testify that it's pretty cold.

But the whole ocean DOES affect our climate. It helps keep us warm in winter and cool in summer. Augusta, Maine, which is under the influence of continental weather, would be colder in winter and warmer in summer. In winter the land is cold and the oceans are relatively warm. In summer the land is warmer than the water.

Augusta is close to the sea, but still gets most of its weather off the land because the air masses in our latitudes most often move from west to east.

The westerly flow of the weather gives Seattle a lot of ocean air in the winter, and that air usually makes its landfall at temperatures above freezing. The Weather Service explains it thusly:

"The prevailing westerly air currents cross vast reaches of ocean, acquiring much water vapor and a temperature near that of the sea. This effect is received from the general currents of the ocean rather than the Japanese Current . . . "

In other words, it's the whole ocean that does the job for

us, not a specific current. So let's bury the old myth and keep it interred.

The Mechanics of Northwest Weather

Highs and Lows

If you could see a high pressure cell in the atmosphere, it would look like Mt. Rainier. And a "low" would take on the appearance of an open-pit mine.

In the northern hemisphere the air flows clockwise around a high, and counterclockwise around a low. Moreover, it tends to flow from a high to a low.

Sometimes two adjacent highs are separated by a trough of low pressure. Two lows, on the other hand, might be kept apart by a ridge of high pressure. A trough would be comparable to a valley between two peaks, whereas a ridge would be similar to a mountain range.

Those are simplified explanations. The whirling earth causes the winds to turn, but topographical features such as mountains and valleys may kick the highs and lows out of shape.

Not only does the air move within each high and low, but the entire systems may move. A high is a bearer of good weather. A low constitutes a disturbance in the atmosphere, sometimes a feeble offering of clouds only, at other times a full gale, tornado, or hurricane.

Many-a-time when the wind sweeps across Washington, it is headed for the Midwest to fill a mammoth low-pressure system. It is flowing from a high over the ocean, and getting a big assist from the prevailing westerlies.

The heat of the sun is the prime mover of our atmosphere. If there were no sun, all the winds of the earth would subside and remain still forever. The chill would deepen, and it wouldn't be long till all life disappeared.

In general (except for complications caused by the turning earth) equatorial air rises, then levels off and heads toward the poles in two different streams. After arriving in the Arctic and Antarctic, the air (then dense and cold) does an about-turn and heads back to the equator. This coming and going is perpetual.

The turning earth's effect upon air is called the coriolis force. The warm equatorial air, flowing north at high levels, begins to pile up at 30 degrees North Latitude due to the revolving globe. Part of the air is turned eastward, becoming a westerly wind; part goes on to the arctic, and part subsides to create and maintain permanent high pressure systems. These subtropical highs in the northern hemisphere are the Pacific High and the Atlantic High.

Now you know why the weather is almost always sunny and warm at 30 degrees North Latitude, and for hundreds of miles north and south of that imaginary line.

The pressure is permanently high at the North Pole, too, but at 60 degrees North Latitude there are permanent low pressure systems, including the well-known Aleutian and Icelandic lows. This area is a battle zone between the polar high and the subtropical highs. It is a source region of miserable weather.

Much of our winter storminess here in the Northwest is literally hurled at us from the Aleutian low, breeder of storms.

Fronts

Take a look at the weather map this morning, and you'll see a front of one kind or another lying over the United States. There might, in fact, be several.

A front is an area of discontinuity between two air masses of differing characteristics. It may be the leading edge of a violent storm, but it could bring only a mild spell of unsettled weather.

Some fronts are warm, others cold. The makeup of each is as different as night from day.

A cold mass of air moving into a region of warm air, and pushing the existing air out of the way, creates a weather disturbance called a cold front. Angry cumulus clouds "boil" upward. A squall line may form, causing heavy rain showers. Or hail may fall as lightning streaks through the tumultuous cloud castles.

A brisk cold front storm is a swifty compared to a warm front. Wintertime cold fronts often move between 25 and 35 miles an hour. They travel more slowly in the spring, but are never slowpokes.

Right now I want to correct a possible misconception. A cold front doesn't always bring chilly weather. This type of weather disturbance could conceivably have temperatures in the 80s or 90s. A cold front is simply the leading edge of an air mass which is COLDER than the air it is replacing. Cold air is heavier than warm air, so the cold front comes in like a mammoth steam roller, rooting under the warm air.

Cold fronts are lively performers. They always put on a good show, and the faster they move, the more spectacular they are. A squall line consisting of towering cumulus clouds, varying in color from white to watery black, is a furious cold front.

Watch for the next cold front. Note the dramatic entrance it makes. And if it's a vigorous storm, be on the lookout for lightning high in the clouds. The accompanying thunder will sound like trunks being moved in your attic.

Now let's examine the warm front. This type of weather disturbance casts out warning signs in the form of high cirrus clouds long before the front itself arrives.

Not all cirrus clouds (the wispy high ones) are banners of a storm, but those that increase and thicken are.

The warm front pulls overhead slowly like a slanting ceiling. It may take a day or more to make its full presence known, whereas a cold front can come and go in a matter of hours.

Because the warm air is lighter than the cold air it is replacing, it actually climbs up the back of the heavier air. That's the reason its leading edge is long and tapering.

Whereas the typical cold front moves swiftly and is characterized by cumulus clouds, the warm front comes slowly and is full of formless stratus clouds.

Warm fronts are seldom as well defined as cold fronts, the surface boundary between the warm and cold air masses being a broad transition zone. The warm front often has an extensive area of rain or drizzle. Both temperature and humidity will climb slowly.

A warm front may produce thunder and lightning, but it's not usually as vicious as a cold front.

The heat-caused lightning storm of summer is not connected with either the classic warm or cold fronts, although it may create some little "fronts" of its own while developing.

Most of those hot-weather storms are born within a single air mass. They develop when a part or parts of the air mass become warmer than the surrounding air. The warmest air soars skyward into a region of lower temperatures.

As the rising air cools, its moisture condenses into cumulus clouds which may grow into gigantic anvil-shaped cumulonimbus clouds. The anvil shape is the tipoff that a thunder-and-lightning "factory" has been born.

The hot-weather thunderstorm is the kind that produces lightning at a fairly low level. The cold front thunderstorm, on the other hand, generally occurs at a high level, and the lightning usually zips from cloud to cloud, and only rarely from cloud to earth.

Clouds—Our Specialty

High in the atmosphere on a summer day, we often see lacy bits of cloud in patches or streamers. Some look like horses' tails or commas; others are in bands, comparable to ripples on a sandy beach. They come in many other shapes and sizes, too.

Those wispy sky riders are made of ice crystals. Their mean elevation is 20,000 feet, but they may be as high as 40,000 feet overhead. If cirrus clouds are few in number and not increasing or lowering, they're merely signs of good weather. However, an old proverb warns of stormy weather if you see certain types of cirrus:

Mackerel sky and mares' tails
Make lofty ships carry low sails.

There are other variations, such as:

Mackerel sky, mackerel sky,
Never long wet and never long dry.

A mackerel sky,
Not 24 hours dry.

Cirrocumulus clouds form the mackerel sky. They look like small, white flakes, usually without noticable shadows, and are arranged in groups or lines. They are uncommon everywhere, and only rarely seen over Seattle.

The "cirrocu" occasionally appear in company with the curly mares' tails, hence the association of the two in proverbs. If they are multiplying, chances are a weather disturbance is on the way. They're way ahead of the storm, perhaps 500 or 1,000 miles in advance of it, because they've been caught in a fast current of wind.

That's why one weather saying tells us that a mackerel sky denotes fair weather for the day, but rain a day or two after. It takes that long for the core of the storm to catch up with its advance banners.

Cirrostratus Clouds

The halos seen at times around sun or moon are the result of light refraction and reflection by the ice crystals of cirrostratus clouds which have the appearance of thin, white veils.

You can look at cirriform clouds and see no apparent movement because they are so far away. If you see them moving, the wind probably is blowing at 70 knots or more at their elevation.

Jet streams of wind in the high atmosphere sometimes hurl cirrus clouds along at speeds of 150 knots.

Halos formed by the passage of light through thin upper clouds are white, as a rule, but some of the rings are red on the inside, shading off to yellow. The ice crystals act in the manner of prisms to break up the solar (or lunar) light into its spectral hues.

While a halo was visible in England, meteorologists flew into the cirrus cloud to take its temperature. The reading at the base of the cloud was 68.8 degrees below zero.

A halo may be a sign of coming rain. Halos are seen most often when clouds are beginning to thicken and lower at the time of a weather change. Several proverbs refer to halos, including these:

When the sun is in his house (halo)
it will rain soon.
The bigger the ring the nearer the wet.
(From the legends of the Zuni Indians.)

The circle of the moon never filled a pond;
the circle of the sun wets the shepherd.

The cirrostratus or veil cloud has a relative called altostratus which lies at a lower level and is thicker. If sunlight is visible through an altostratus layer it appears to be coming through ground glass and shadows are not cast by the light.

There's another difference. Cirrostratus clouds are made of ice crystals, whereas altostratus may be all water or a combination of water droplets and ice.

It's not uncommon for water droplets to exist, unfrozen, in clouds whose temperature is below 32 degrees. Water has been observed in fog having temperatures as low as 40 below zero.

Such supercooled water seems to be dependent upon an undisturbed condition of the molecules in it. Agitation of the air, or contact of the water with a solid surface, such as an ice crystal, causes instant solidification.

The cumulus tufts in the cirrus family are 32d cousins of the better-known cumuli we see throughout the summer. These low cumuli are clouds of vertical development. Some are small and benign, looking for all the world like heaps of soft cotton; they are known simply as cumulus of fine weather. You've probably heard the saying:

When clouds appear like rocks and towers,
The earth's refreshed by frequent showers.

Those are cumulus that shoot upward in unstable air. They often produce showers only, but may continue rising until they achieve the anvil shape of thunderclouds.

The thunder-and-lightning cloud is given a name of its own, cumulonimbus. Its bottom may lie somewhere around the 1,600-foot level while its top is pushing into the tropopause, the zone that separates the dense, lower atmosphere from the cold, quiet atmosphere.

There are many shapes and sizes of so-called middle clouds that lie somewhere around the 10,000-foot level. Some of these clouds, called altocumulus and altostratus, may drift lazily on summer days, but at times they lower and become rain bearers.

In general, you can expect good weather if the high cirrus and/or the middle clouds are decreasing, or at least not increasing and lowering. But if they're dropping and thickening, look out!

The warning is particularly applicable in the winter, spring and fall when our storms are more intense than those of summer.

Raindrops—A Close-up

Dust helps make the sky blue. It helps make rain, too.

The sun's rays are scattered by water vapor, molecules of dry air, and solid particles like dust. The shorter the wavelength of the light, the more it is scattered. Blue is scattered the most and red the least; that's why we look upward into a blue vault on clear days.

In the early morning and late in the day, we are likely to see more red and pink in the sky because the sun's light has to travel a greater distance through the atmosphere at those times. This difference in the slant of the sunbeams causes a loss of much blue light and accentuation of the red and makes many sunrises and sunsets spectacular.

It is this phenomenon, however, which also caused seafarers to coin the phrase:

Red sky at night is a sailor's delight;
Red sky in the morning, sailor take warning.

Every raindrop needs a nucleus in order to develop. Dust is one of the main sources of these nuclei. Some of the dust is blown into the air from the ground by the wind. Other producers of atmospheric dust are meteoric showers and volcanic eruptions.

As the earth swings around the sun in its orbit, it runs into swarms of meteors at certain times of the year. Many of the meteoric fragments are so tiny that it's difficult to see them, even with a microscope.

It has been noted by meteorologists that heavy rainfall occurs over many parts of the earth about 30 days after a meteor shower. A series of such storms has been noted, particularly in January—usually about the 12th, 22d, and 31st—following the earth's passage in December through areas where meteors are concentrated. Some of the nuclei for raindrops may have drifted to earth from comet tails.

Gigantic eruptions of volcanos are almost always followed by months (or years) of hazy weather around the world due to the dispersal of volcanic dust.

It is a matter of record that the eruption of Krakatoa, a volcanic island in Sunda Strait, halfway between Sumatra and Java, in August, 1883, hurled so much dust aloft that the sun was either totally or partially obscured for weeks in various parts of both hemispheres.

Generous rainfall for months following the Krakatoa eruption was attributed to the presence in the air of so much of Krakatoa's dust.

It is probable that the Krakatoa explosion was the greatest that ever occurred on earth. Much of the island was blown away and a tidal wave washed over nearby islands, killing 30,000 persons. The dust traveled around the world several times before settling.

The weather of 1816, which was cold and wet in America even during the summer, has been blamed on volcanic eruptions in various parts of the world prior to 1816. That was the year in which freezes occurred and snow fell during each of the warm months.

Raindrops are portrayed most often by artists as tear-shaped, looking somewhat like a miniature pear. Fact is, raindrops can change their form as often as 50 times per second, according to General Electric Company scientists.

At times the falling drops look like jelly beans, but before they reach the ground they may resemble for fleeting moments such things as pancakes, gourds, peanuts, hotdogs, ducks, footballs, and even human feet.

To study raindrop behavior, GE scientists developed an instrument called a drop controller which produced air resistance similar to that encountered by falling drops in the free air.

By regulating an upward air current, the rate of fall of the drops could be speeded up, slowed down, or halted, thus simulating conditions encountered by raindrops tumbling from a cloud. The changing shapes were photographed with highspeed cameras.

The Rough Stuff—Earthquakes, Tornadoes, Hurricanes

Is there such a thing as earthquake weather?

The answer is "no." Science hasn't yet found any evidence that a specific kind of weather—wet or dry, hot or cold—is associated with quakes. Rocks creep and slip deep in the ground to cause temblors that shake the earth from time to time, but these movements occur in all kinds of weather.

However, there is one factor, air pressure, which MIGHT have an effect. A sharp rise or fall in the pressure (or weight) of the air conceivably could be enough to trigger a quake if other conditions were favorable.

The late Frank Neumann, one of America's leading seismologists, told me that changing pressure, by itself, isn't likely to set off a major earthquake, "but it could be the one straw needed to light the fuse."

Weather didn't cause the calamitous San Francisco earthquake of 1906, but it can be held responsible for much of the devastation by fire that followed.

The quake occurred at 5:13 a.m. on Wednesday, April 18, caused by a movement along a 200-mile section of the famed San Andreas Fault. It was felt over an area of approximately 375,000 square miles. The earth was shaken from southern California to northern Oregon.

On that fateful morning, San Francisco was getting a flow of hot, dry air from the northeast. No rain had fallen since April 1. To make matters worse, the quake broke the water mains.

The fire spread quickly in the warm air. A gentle breeze fanned it westward through the business district and into residential areas. The fire burned furiously on the 18th, 19th and 20th. There were isolated blazes that lasted longer.

The heat spell began on the day of the quake and didn't end until Saturday night, April 21, when a half-inch of rain fell over the bay region.

If San Francisco's normal April weather had prevailed, the fire might have been confined to the city's lower business areas and might have burned out along the bay shore waterfront.

The maximum temperature at Oakland on the day of

the San Francisco earthquake was 70 degrees Fahrenheit.

San Francisco's data isn't available because weather instruments there were destroyed by the quake.

The loss in San Francisco due to the earthquake has been estimated between 5 and 25 percent. Ten percent was the figure given by most investigators, according to the Seismological Society of America.

That means most of the damage—estimated between $350 and $450 million—was caused by fire. The weather was an accessory after the fact, and is guilty as charged.

Quake Weather? I Repeat, There Isn't Any

There's a widespread belief that quakes occur when the air is calm, hot, and muggy, and I have no doubt that many temblors occurred when those conditions prevailed. But it takes more than coincidence to make a rule.

Now let's have a look at Seattle's weather records for the dates of our two worst temblors, April 13, 1949, and April 29, 1965. I also have a report from the National Weather Service office at Anchorage concerning conditions that prevailed at the time of the great Alaska Earthquake of March 27, 1964, one of the most destructive shocks in the history of the world.

Seattle's shakiest temblor within the memory of anyone alive occurred at 11:55:41 a.m. on April 13, 1949. At that moment the barometric pressure was 30.26 inches, the sky nearly overcast, the temperature 50 degrees, and the wind four miles per hour from the north.

A weak high-pressure system was centered in the Pacific Ocean off Vancouver Island. The remnants of a minor storm had moved through the city the night before, but by earthquake time the storm was in the Cascade Mountains.

The sky was completely overcast up to 7:30 a.m. on the 13th. From then on there were breaks in the overcast. Between 11 and 12 noon clouds covered a little more than nine-tenths of the sky.

The barometric pressure varied but little between 8:30 a.m. and 2:30 p.m. on the 13th, and there was no change whatsoever during the quake. The pressure reading at 8:30

a.m. was 30.24 inches; at 9:30 it was 30.25; at 10:30, 11:30, 12:30 and 1:30 the pressure was 30.26 and at 2:30 the reading was 30.25.

The statistics for that date speak eloquently. It was a nice spring day, neither muggy nor oppressive.

That quake, by the way, registered Number VIII (in some areas) on the Modified Mercalli Scale. A temblor achieves Number VIII if damage is slight in specially designed structures, considerable in ordinary substantial buildings, and great in poorly built structures.

Our earthquake on April 29, 1965, registered Number VI on the Mercalli Scale. It occurred at 8:28:54 a.m. when the temperature was 47 degrees, the sky clear, the wind east-southeast at four miles an hour, and the pressure 30.14 inches.

A ridge of high pressure was influencing Seattle's weather at the time. The only weather disturbance of any consequence lay more than 600 miles to the west-northwest, and it showed little disposition to move fast.

The relative humidity was high—80 percent—at earthquake time, but the temperature was low enough to eliminate any hint of mugginess.

Let's backtrack now to 5:36 p.m. on March 27, 1964, the time and date of Alaska's devastating quake. In Anchorage, the state's largest city and one of the places hardest hit, the temperature was 28 degrees Fahrenheit and the barometric pressure 29.91 inches. Skies were overcast, and a light snow was falling. The wind was from the northwest at four miles an hour.

While there's no such thing as "Earthquake Weather," tornadoes and hurricanes *definitely* have a season.

Most tornadoes in this country occur on hot, sticky days in April, May, and June. Texas is the state most frequently hit by this type of storm, the most violent on earth, yet every state has reported funnel-cloud twisters.

The all-time record for tornadoes in a single calendar year in the United States was broken in 1965 when 927 were reported. That's almost 50 percent more than the average per year, 628, for the period from 1955 to 1964, inclusive, according to the National Weather Service.

The second worst tornado year was 1957, with 864 twisters, and the third worst, 1964 with 722.

In Texas the average number of tornadoes per year is 28, but the 1965 total there was a whoppin' 108. Neighboring Oklahoma was in second place in 1965 with 74, in contrast to an average of 24, and Kansas had 69 (normal 26).

South Dakota was hard hit, too, in 1965 with 64 twisters (normal five), and Indiana had 48 (normal seven).

One of the most surprising totals, 41, was chalked up in Colorado, a state that isn't even part of America's Tornado Alley. Colorado's average is four funnel storms per year.

The states normally experiencing the fewest tornadoes are those in the Far West and Northeast, and the twisters that develop in those regions are weaklings compared to those in the lower Midwest.

Washington's average is less than one per year. Other states with fewer than one are Oregon, Idaho, Nevada, Utah, and Arizona.

California's average is one, and Montana's, Wyoming's, and New Mexico's two each.

Tornadoes caused more than four times as many deaths and injuries in 1965 as in 1964. Most of the 1965 twisters occurred in May, as usual, but those of April, 1965, sent damage figures soaring to an all-time high for one month.

The tornadoes of April, 1965—47 of which occurred on the 11th, Palm Sunday—claimed 257 lives and caused damage estimated in the neighborhood of $1 billion.

The death toll in that Palm Sunday tornado outbreak was exceeded only by a comparable series of twisters on March 18, 1925, when 689 persons were killed and 1,980 injured in Missouri, Illinois, and Indiana.

Damages resulting from the Palm Sunday storms were greater than for any full year of tornado activity in this country since 1915.

The Palm Sunday tornadoes cut a swath of destruction through Iowa, Illinois, Indiana, Ohio, Wisconsin, and Michigan. They formed when warm, moist air from the Gulf of Mexico met a fast-traveling stream of cool, dry air out of the west. Thunderstorms developed along the boundary between the two air masses.

It is not uncommon for two such air masses to collide and spawn thunder, lightning, and a few tornadoes in the springtime. But in the case of the Palm Sunday storms, the jet stream became a contributing culprit.

The jet stream is a fast-flowing current of air at high levels in the atmosphere.

On that particular day the jet stream entered the United States from the west over Southern California and flowed northeastward toward New England at speeds ranging from 140 to 185 miles per hour. Its path took it over Arizona, northwest New Mexico, southeast Colorado, Kansas, Missouri, Illinois, and on to the east.

In the Far West the high-level air stream caused no special turmoil, but as it entered the Midwest, it stirred the already turbulent air where thunderstorms had begun to form. That was the spark that triggered the writhing tornado funnels, one after another across the plains.

Each of the 47 twisters lashed at part of the continent's mid-section with winds blowing hundreds of miles an hour within the funnel.

Many areas escaped the fury of the Palm Sunday tornadoes but suffered damage due to hail, lightning, or flash-floods.

Fourteen of the tornadoes traveled together, like a family, through 24 counties in Indiana, causing damage in that state along in excess of $100 million. That "family" of funnels destroyed 1,250 houses, 221 house trailers, and 922 other buildings in Indiana, and damaged thousands of other structures.

The most vicious of the Palm Sunday twisters probably was the one that swirled through Branch, Hillsdale, Lenawee, and Monroe counties in Michigan. It killed 44 persons, injured 612 others, and hit 1,590 homes. Thousands of automobiles and boats also were destroyed or damaged. Michigan's estimated damage was $500 million.

The Far West has never had a tornado comparable to any of the Palm Sunday storms of 1965 because we don't have violent clashes between vastly different air masses of the type that meet head on beyond the Rockies.

Our mountains are a help, too. Tornadoes find the going rough in hilly country.

September is hurricane-time in the Caribbean Sea, the Gulf of Mexico and a large part of the North Atlantic Ocean.

Thirty-three percent of nearly 300 hurricanes over a period of 45 years occurred in September. Twenty-seven percent developed in October and 18 percent in August. The others (22 percent) were spread out over the rest of the year.

The frequency of these storms is directly related to the warmth of the air and water in the tropics. The greater the heat, the more likelihood of hurricanes developing.

The highest temperatures occur over the water near the equator after the summer solstice (June 21). The build-up of heat occurs over a period of months, which accounts for the long hurricane season.

Microclimates

Backyard Weather

How's the microclimate in your yard? Maybe you weren't aware of it, but you have a small weather pattern of your own, and it might be quite different from that of your neighbors.

This personal climate exists in a small area. It is affected by your lawn, shrubs, trees, flowers, fence, and the shape of your house. You can change it by planting a shrub or digging one up.

The micro part of the microclimate means "small in size." The American Meteorological Society's "Glossary of Meteorology" says the microclimate varies with and in turn is superimposed upon the larger scale conditions. And there's a parenthetical note saying: "Observe the microclimate of a putting green versus that of a redwood forest."

Generally, four times the height of surface growth or structures defines the level where microclimate overtones disappear.

A line of trees in the general vicinity of your home, plus

the shade of a spreading Douglas fir in your yard might be just enough to make your place pleasantly comfortable on a hot day while your nearest neighbor had the equivalent of a climatic bake oven.

Man is constantly seeking a preferred microclimate for himself. He heats the house where he lives and the building in which he works. He seeks the shade of a friendly tree on a hot day, or a spot in the sun if he's at the beach on a day when the water and a fresh breeze have cooled him to the shivering point.

Farmers live closer to nature's unmolested microclimates than anyone else, except for natives on remote islands, hermits in the hills, and workers in the deep forests. Yet farmers do more, deliberately, to alter the large-scale climate and create a microclimate than city dwellers.

I refer to the planting of trees in rows to subdue the wind. Such windbreaks not only prevent grains and other limber growth from being leveled by a furious blow, but they help produce greater yields on farms, in orchards, and on ranches where cattle and other animals are raised.

The most effective windbreak in windy country consists of several rows of different species of trees. Five rows is ideal in most cases, although many good windbreaks contain fewer than five rows.

The ideal setup for a five-row windbreak would be a lineup like this, starting from the windward side: row 1, a dense shrub, deciduous; row 2, a medium-sized tree, deciduous; row 3 (in the middle) a very tall tree, deciduous; row 4, a tall evergreen; row 5, a medium-sized evergreen.

A dense, full-grown windbreak consisting of several rows of trees will reduce the wind speed for a distance of about 30 times the heighth of the tallest row.

Some Lakes Really Flip

Every man should have a small lake in his life. And a brook. And time enough to sit beside each, once or more each season, to calm the nerves, refresh the soul, and listen to nature's small talk.

Henry David Thoreau, naturalist and writer, went to

Walden Pond at Concord, Massachusetts, to meditate upon the basic truths of life while tuning in on nature's rhythms.

I'm not suggesting that you flee from family, job, and freeway traffic, but you can and should seek out nature once in a while to partake of her medicine for frazzled nerves. The tonic is free and the benefits are long-lasting.

Mention of Walden Pond reminds me that most small lakes and ponds (and many big ones) undergo an overturning twice each year. The overturns, occurring in the spring and fall, give the lakes new "life" and vigor. In contrast, stagnation occurs in the summer and winter.

It might be said that each lake takes a deep breath each spring and fall. The factors involved are temperature, gravity, and wind.

I won't bore you with statistics, but there's an important figure to keep in mind, 39.3 degrees Fahrenheit. Water is densest at that temperature Fahrenheit.

Dense water, like dense air, sinks to the lowest level it can find. So the densest water in any lake is going to be at the bottom. And its temperature is likely to be 39.2 degrees, or close to that figure, all year long.

Now let's create a mythical lake to see what goes on. We'll give this body of water a surface diameter of 1,500 feet and a depth of 200 feet.

In the heat of the summer, while the bottom water has a temperature of 39.2, the surface is likely to warm up to 65, 70, 75 or even higher. And the water will be warm down to a depth of 20 or more feet, with very little loss of heat in that upper layer.

Scientists call such a warm layer of water the "upper lake" in contrast to the dense, cold "lower lake." In between the two is a transition zone known as the thermocline in which temperatures drop sharply.

The result is stratification of the lake into two distinct microclimates, which prevents thorough mixing. The only circulation of any consequence is that caused by summer breezes which might turn over the water in the warm upper layer, but wouldn't disturb the "bottom lake."

The "bottom lake" under such conditions would stagnate and become oxygen-poor. Fish and other lake life able to move away (and up) would shun it.

Now let's have a look at our mythical lake in winter. The cold, dense water is still at the bottom. The temperature there is still at, or close to, 39.2 degrees.

Remember, water is densest at 39.2 degrees. If it cools to a lower reading, it becomes lighter. So the lightest water in the coldest part of the year is on top, perhaps at 33, 34, 35, or 36 degrees, but sometimes at 32, the freezing point. In its solid form (as ice) water is unusually light.

In analyzing our mythical lake, we find the densest water at the bottom and the lightest on top, both summer and winter.

How, then, is an overturn achieved? Here's how:

After winter's back is broken, the lake surface starts warming up. Let's say the surface temperature on March 1 was 34 degrees. A little later it heats to 35, then 36, then 37, and so on.

Finally the magic number, 39.2 degrees, is reached. Suddenly there are no more thermal layers. The whole lake is in equilibrium. Then a vigorous wind springs up. The lake waters begin to roll. The surface waters spin downward, and the bottom waters come to the surface.

The lake is turning over, taking its deep breath. New oxygen is whipped into the water by the wind and carried from top to bottom.

The lake's slow roll is imperceptible to the eye, but you will be aware of the occurrence if you see decomposed matter or bottom silt suddenly appearing on the surface. Such rising of bottom material occurs more often in the fall than in spring.

The turnover occurs again in the fall when the surface temperature gradually drops into the 50s, then the 40s, and finally reaches the magic number, 39.2 degrees.

There is little resistance to mixing during the overturn periods, but considerable resistance in summer and winter, due to the existence of heavy and more viscous (less fluid) water in the lake bottom.

The lower lake becomes oxygen-poor twice a year

because decomposing material at the bottom causes a continuing drain upon the oxygen supply, and because no generous new supplies of oxygen are available, except during overturns.

If I ever become a lake fisherman (I prefer the rivers), I'm going to remember NOT to fish deep-down unless I'm sure that underwater springs or incoming streams are carrying oxygen to the depths both summer and winter.

Fish need oxygen, even as you and I.

The Northwest's "Superb" Climate Compared

I DIDN'T know Seattle ever had any pleasant, rainless weather like this. Why aren't you telling the world about it?"

You've heard that comment, either from convention visitors or your Aunt Minnie from Green Bay, and it surprised you because you thought most everybody knew about our glorious summers and moderate winters.

Maybe we haven't been diligent enough in "selling" our climate to those who swelter in summer and freeze in winter. So here are some statistics on the weather here and elsewhere.

Northwest weather is far less severe than it is in nearly all places across the nation in our latitude, which is 47 degrees, 36 minutes, 32 seconds North. We are warmer, too, than many places farther south, thanks to the moderating effect of the Pacific Ocean, plus the protection our mountains give us from icy winter blasts.

Seattle's average annual temperature is 51.1 degrees Fahrenheit. The averages for our warmest and coldest months—July and January, respectively—are 64.5 and 38.2. Normal precipitation is 38.8 inches.

Compare the Northwest's moderate temperatures with these figures from other parts of the country:

Portland, Maine (43 degrees 39 minutes North Latitude)—Warmer than Seattle in summer, colder in winter. Mean annual temperature, 45 degrees Fahrenheit, six degrees below Seattle's, despite the city's position farther south. January's mean temperature, 21.5 degrees, is nearly

17 degrees chillier than the January figure for Seattle. July's mean temperature, 68 degrees, exceeds Seattle's July average by 3.5 degrees. Portland's snowfall averages 74.2 inches in a year's time, and the snow is on the ground for most of the winter!

Boston, Massachusetts (42 degrees, 22 minutes North Latitude)—Considerably warmer than Seattle in summer, colder in winter. The extremes result in a mean annual temperature of 51.3 degrees, almost the same as Seattle's. The mean January temperature is 29.2 degrees, nine degrees below Seattle's figure for the month; mean July temperature, 73.3 degrees, which is 8.8 degrees above Seattle's July average. Boston's snowfall averages 42 inches per year in contrast to Seattle's 14.8 inches.

New York City (40 degrees, 47 minutes North Latitude)—Summer's extremes of heat and winter's extremes of cold are far greater than Seattle's. The mean temperatures for January and July, respectively, are 32.2 degrees and 76.6 degrees, compared to Seattle's 38.2 and 64.5 for those months. The warmth comes in May and lasts far into September, except for customary—and welcomed—relief that comes with cooling breezes.

However, summertime mugginess is a characteristic of East Coast weather, and New York isn't exempt. The mean annual temperature is 54.5 degrees; mean annual precipitation, 40.19 inches, and mean annual snowfall, 29 inches.

The city has had a temperature reading as low as 13 degrees below zero in December and 15 below zero in January. The all-time high was 106 degrees one July day in 1936.

Rapid City, South Dakota (44 degrees 3 minutes North Latitude—Blazing hot in summer, icebox weather during many winters. Mean temperatures for the coldest and warmest months—January and July—are 21.9 and 72.6 degrees Fahrenheit, respectively. The average daily minimum temperature in January is 9.6 degrees (that's mighty cold, brother!), and July's average maximum is 86.3. The mean annual precipitation is 17.12 inches. Snowfall averages 38.7 inches per year.

What's your choice, Rapid City or Seattle? I'll take our Puget Sound climate!

Washington, D.C. (38 degrees 51 minutes North Latitude)—Summers are warm and humid, winters reasonably mild. Generally pleasant weather prevails in the spring and fall. The coldest weather occurs in late January and early February; the warmest occurs late in July. The mean annual precipitation figures for Seattle and Washington are almost the same (our 38.79 inches, the capital's 38.89), and there isn't much difference between snowfall totals (our 14.8 inches, Washington's 16.4).

But the statistics can be misleading in making comparisons. We can chalk up a plus for our climate because of low relative humidity in the summer, whereas the nation's capital gets a minus in my book because of summer's moist air. Heat and high humidity are a wearying combination.

Atlanta, Georgia (33 degrees 39 minutes North Latitude)—Rainfall is generous at all seasons in Atlanta, ranging from a mean of 2.5 inches in October to nearly 6.0 inches in March. The average annual snowfall is about 1.5 inches, although a snowfall of 4.0 inches or more is likely to occur about once in five years. The record snowfall, 8.3 inches, occurred January 23, 1940.

Mean annual precipitation totals 48.34 inches, and a lot of that moisture falls during summer's thunderstorms, which are numerous. July, the warmest month, with a mean temperature of 78 degrees, normally has about 11 thunderstorms. The summer weather is muggy too, but prolonged periods of hot weather are rare.

One advantage that Atlanta has over Seattle is more sunshine. The number of clear or only partly cloudy days exceeds the number of cloudy days.

Miami, Florida (25 degrees 48 minutes North Latitude)—Nature is generous in providing Miami and the surrounding area with plenty of water to drink and nurture the region's crops. The climate is defined by the National Weather Service as "subtropical marine," featuring long, warm summers and abundant precipitation. The annual rainfall averages more than 59 inches. Mean monthly temperatures exceed 67 degrees in winter and 80 degrees in the summer.

The city lies in a belt of frequent thunderstorms, most

of which occur in summer. There are, on the average, 7 such storms in May, 13 in June, 15 in July, 16 in August, and 12 in September.

Miami's mean maximum temperatures in summer reveal why citrus and other warm-weather fruits do well in that part of the United States. The average daily maximums in summer are 88 degrees in June, 89 in July, 89.9 in August, and 88.3 degrees in September. Even October's daily maximum is high, 84.6 degrees.

Freezing temperatures have occurred in the farming districts southwest, west, and northwest of Miami, but almost never near the ocean.

Even if you leave the United States, you're not apt to find a better climate than right here in the Northwest. Here are some international weather figures—decide for yourself:

Kobe, Japan, our sister city, is warmer and sunnier than Seattle, and wetter too. Seattle's average annual temperature, 51.1 degrees, compared to 59.5 for Kobe. Seattle's annual precipitation, 38.79 inches; Kobe's, 52.63.

Seattle is comparatively dry in summer, whereas Kobe normally gets a little more than seven inches of rain in June, about the same amount in July, and more than five inches in August.

Seattle and Kobe have had the same all-time high temperature, 100 degrees. Our all-time low was zero on January 31, 1893; Kobe's lowest reading was 20 on January 18, 1936.

Seattle's average annual snowfall, 14.8 inches; Kobe's less than four inches. Our heaviest snowfall in 24 hours, 21.5 inches in February, 1916; Kobe's deepest snow was seven inches in February, 1945.

London, England (51 degrees 30 minutes North Latitude)—Although four degrees farther north than Seattle, London's weather is strikingly similar to ours. The air throughout Britain has more of a maritime flavor, however, especially in summer, because the country is an island.

Average temperatures are comparable to ours, month-for-month, but London is drier, with a mean annual rainfall of less than 30 inches. London's cloudiness is similar to ours, too, especially in the fall, winter, and spring.

As for fog, it's a toss-up, although London's fog is browner than ours—and perhaps more dense—indicating the presence of more pollutants in England's air on foggy days.

Berlin, Germany (52 degrees 32 minutes North Latitude)—The city, situated approximately 225 miles from the nearest ocean (the North Sea), has a continental-type climate similar to the climate of many inland cities on our continent.

Summers are similar to Seattle's, but winters are colder than ours. July's average temperature for Berlin and Seattle is precisely the same, 64.5 degrees. January, the coldest month for both cities, is chillier in Berlin. January's mean temperatures are 31 for the German city, 38.2 for Seattle.

As for umbrella weather, Seattle wins (or should it be loses?). The mean annual precipitation for Berlin is less than 25 inches.

Paris (48 degrees 50 minutes 14 seconds North Latitude)—Parisian weather records, kept for well over 150 years, show that winters are a bit colder than Seattle's, summers are warmer, and the whole year is drier. Mean January temperatures are 37 degrees for Paris, 38.2 for Seattle. Means for July, the warmest month, are 65.5 for Paris, 64.5 for Seattle.

Parisian rainfall data (a continuous record for nearly 140 years) show that precipitation doesn't differ greatly from season to season, although the summer months are slightly wetter than winter months. The mean precipitation per year is only 22.3 inches. (Note that Paris lies slightly north of Seattle's parallel of latitude.)

All in all, looks like we're better off at home!

Seattle's Shower Power and Second-Rate Storms

Northwest weather fans will grudgingly admit that, "It gets a little moist around here in the fall, winter, and early spring," but I didn't realize just how much tonnage piled up in the form of rain until I wrestled with some figures:

One inch of rain covering the entire city of Seattle (91.6 square miles) would weigh 6,627,761.38128 short tons (2,-000 pounds per ton). That figure multiplied by the city's mean annual precipitation, 38.79 inches, gives a total of 257,090,863.98 tons in a year's time.

Man, oh man! The Sahara Desert could use some of that moisture couldn't it? So could Arizona and Southern California.

The water in just one inch of rain over Seattle's 58,624 acres would total 1,591,892,845 gallons. That amount would be enough to satisfy all customers of the Seattle Water Department for a few days, even in the summer when lawns and gardens are irrigated.

Seattle's mean annual precipitation is recorded at Seattle-Tacoma International Airport, where the National Weather Service takes observations. The airport records became the official climatological data for Seattle after the downtown forecasting center moved to the Lake Union Building in 1972.

The average person in our temperate zone needs approximately five and one half pints of water each day if he/she is moderately active. We drink about three pints of fluids daily. The rest of the fluid supply is taken in with the food we eat, or created in the body by oxidation of food, principally sugar, starches, and fat.

The air itself contains water, even when no rain is falling. This invisible water vapor is required for breathing and to keep our skin healthy.

Digestion can't take place without water either. Nor could our body dissipate heat on hot days if it weren't for water exuded by the skin as perspiration.

The United States Geological Survey estimates that 4,300 billion gallons of water fall daily upon the nation in the form of precipitation. The average annual precipitation is 30 inches.

Farmers who irrigate their crops are the principal users of fresh water in the United States, requiring in excess of 100 billion gallons daily, or nearly half of the fresh water used annually.

Industry is the second largest consumer of water. A large paper mill, for example, uses more water than a city of 50,000 inhabitants.

The present demand on fresh water supplies above and below the ground in this country per year exceeds 63,000 billion gallons.

It's rather obvious that life itself depends on water. Now if we could induce nature to distribute the commodity a bit more evenly, we'd have a better world.

We're fortunate here that the mountains are high enough to gather and hold a substantial winter snowpack. That snow is our guarantee of a glass of water—cool, clear water—on the first hot summer day.

In spite of our tons of rain, we don't usually get severe storms. In fact, thunderstorms in the Puget Sound region are strictly second-class compared to those over the Great Plains, in the Deep South, and along parts of the Atlantic Coast.

Our lightning is less spectacular too, and our hail is inferior. But no one is complaining.

Even our rains, generous though they are in the late fall, winter, and early spring, are gentle in contrast to some of the downpours beyond the Rocky Mountains. An aviation forecaster for the National Weather Service here, summed it up nicely when he commented: "Seattle people don't know what a heavy rain is."

There is less risk of hail damage in Washington than in any other state west of the Mississippi River, excepting Nevada. Reasons: (1) most of the hail here is small, especially in Western Washington, and (2) much of our hail falls in late winter and early spring before new growth has reached the point where hail is a menace. The worst hailstorms in Washington, with damage to soft fruits and other crops, have occurred in Chelan, Yakima, Douglas, and Columbia Counties, all on the other side of the Cascade Mountains.

Hail forms most readily when strong up-and-down drafts exist in the atmosphere. Such conditions are most prevalent here in the transition period between winter and spring.

Most of our hail in this area is from one-eighth to a quarter-inch in diameter. It gets a little bigger in Eastern Washington.

A pilot flying on the east side of the Cascades reported

seeing hailstones three-quarters of an inch in diameter. That's unusual, though, for Washington.

Some of our so-called hail is "graupel," also called snow pellets or soft hail. Graupel looks like crumbly pellets of snow—miniature snowballs—and ranges in size from that of buckshot to small peas. Graupel is whiter than hail.

True hail has a dull, icy look. Hailstones grow by the accretion of water upon ice crystals. They are a product of cumulonimbus clouds—the kind that create thunder and lightning, as well as tornadoes—and reach their greatest size where atmospheric fury is most pronounced.

The same atmospheric fury that creates big hailstones is responsible for tornadoes and spectacular lightning. You have to see a Midwest lightning display to realize how pale ours are in comparison.

The cumulonimbus clouds in which thunderstorms are born may rise to heights greater than ten miles. The storm is generated by unstable air, moisture, and a lifting force called convection. The updrafts (convective currents) carry warm air high into the freezing atmosphere.

Such storms, which develop entirely within one cloud, contain both updrafts and downdrafts, plus moisture, plus negative and positive charges of electricity. The result is wild confusion.

One cumulonimbus cloud is capable of producing its own fury in the form of heavy rain, hail, lightning, and thunder. But at times two or more of the clouds will develop into a cluster, each putting on its own spectacular show. When that happens, it's an awesome sight.

Each flash of lightning is a gigantic spark between areas where positive and negative charges of electricity are concentrated.

Lightning is one of nature's prettiest creations; I enjoy such a show, but the big thunderstorms just don't often happen here.

Weather and Your Health

HOW would you like to have sunny weather every day, with never a wild gale, no cold waves, and just enough rain to keep the grass green?

You would? Well, banish the thought! You are peppier and healthier here, where the weather is changeable and somewhat unpredictable, than you'd be in a hot country.

The world's most energetic people live in air-conditioned regions like ours where we are likely to get roughed up a little by the elements.

The temperate zones of the earth, which are anything but temperate, are the world's energy zones. And the best parts of these temperate zones have stimulating winters and summers which aren't sultry.

The United States is fortunate; most of its weather is invigorating. Yet some of the states have debilitating weather part of each year. But enervating conditions are rare in our Northwest.

Extremely high temperatures make most people irritable and incapable of doing their best work, either mentally or physically. Students who take tests in cool, changeable weather will get better grades than on a hot day. The best results, mentally, seem to be achieved in the spring and the fall in the northern tier of states.

Experiments with rats at the University of Cincinnati illustrate the point. One group of rats was kept in quarters where the temperature fluctuated between 65 and 70 degrees; a second group lived in a man-made climate comparable to the tropics. The cool rats in Group 1 were robust and healthy with considerable immunity to disease, whereas the rodents of Group 2 grew more slowly, matured later in life, were less fertile, and had weaker, smaller offspring than their colder cousins. But a change of climate proved beneficial for the warm rats; their vigor increased with a decrease of temperature.

The effect of heat on the mental responses of rats also was studied at the University of Cincinnati. One group required an average of 12 tries before finding the way through a confusing network of paths and passages while the temperature stood at 65 degrees. A second group, tested with the mercury at 76 chalked up an average of 28 trials and errors before reaching the end of the maze. Another group of rats was sent through the obstacle course when the thermometer registered 90 degrees and the showing was pitifully poor, an average of 48 trips being required before the problem was solved.

Mental activity for humans reaches a maximum when the outside temperature averages around 38 degrees, perhaps dropping low enough at night to leave a touch of frost on the ground. That's the kind of weather Seattle experiences during much of the late fall, winter, and early spring.

Not Too Hot, Too Cold

Those who work in offices and factories seem to do their best work if the temperature is slightly under 70 degrees, provided there are no disconcerting drafts.

When the temperature reaches 40 degrees below zero, almost all of man's energy is devoted to survival measures, and no useful work is achieved.

The relationship of temperature and humidity is extremely important in our lives. For example, most of us experience no discomfort when the relative humidity is 100 percent, provided the temperature is below 71.

The comfort zone for humans continues according to the following scale:

If the temperature is 80, the relative humidity should be below 66; if the temperature reaches 90, the relative humidity should be below 43, and at 100 degrees the relative humidity ought to be seven or below.

The relative humidity (expressed in percent) is the amount of water vapor in the air compared to the amount it could hold at a given temperature.

All of us humans are affected by the temperature night and day, every minute of our lives. Shivering is one of the bodily reactions that helps warm us on a nippy day. When it's hot, we lose heat by perspiring.

We not only think slower when it's too hot, excessive heat also leads to crimes of violence. A study of 40,000 cases of assault and battery in New York City showed those crimes increasing from cold January to warm July in almost the same proportion as the rise in the temperature.

The movement of air is a major factor affecting our health and well-being. Western Washington is lucky in having a new and fresh supply of air every few days, and sometimes every day; in fact, sometimes every few hours. The prevailing westerlies do a good job of cleaning out industrial impurities in our atmosphere and keeping smog from forming.

There are many less-fortunate parts of the United States and the world where impurities pile up in stagnant air. Some tropical lands have endemic diseases because of stagnant air which encourages disease germs to multiply.

The bitter cold at the top and bottom of the earth, and the fiery heat of the tropics, are the worst for mankind. The weather near the poles seems to be the hardest to take for long periods of time, which accounts for the sparse settlements in those parts of the world.

Caucasians have been going into the tropics for hundreds of years to work, fight, or just enjoy the sun. But they've had to go back to their cooler homelands from time to time in order to regain their physical and mental vigor.

Caucasians also are inclined to have more illnesses and more accidents in the tropics than in colder climes. Their morale is usually lower, too, in the hot belt near the equator.

Humans aren't the only ones who thrive in some climates and suffer in others—so do plants and animals.

Some interesting temperature information has been compiled by the Agricultural Experiment Station, University of Minnesota. The comfort zone for cows of European origin—Holsteins and Jerseys—lies between 30 and 60 degrees, the station reports. But the comfort zone for Brahman cattle (originating in India) lies between 50 and 80 degrees. And Santa Gertrudis cattle (a genetic combination of Brahman and Shorthorn breeds) can tolerate almost as much heat as the Brahmans but are able to withstand much more cold.

The common house fly is most active at temperatures of 60 and above, and the honey bee is active between 50 and 95 degrees. Bees die from heat at temperatures between 115 and 118 and from the cold at 28 to 30 degrees.

Potatoes and garden peas grow best, according to the Minnesota tests, between 55 and 65 degrees, turnips, carrots, cauliflower and celery between 60 and 70, and cucumbers between 65 and 75.

Tomatoes experience their optimum growth between 70 and 75 degrees, but have a tolerance range extending from 65 to 80 degrees for good development.

Things to Know

Moving to another climate isn't likely to help avoid colds and other respiratory diseases. These ailments seem to be about equally common in all regions. Doctors point out that we get used to the climate of our area and quite often are better off there than elsewhere, except for vacation trips which might prove remarkably beneficial.

The brief trip away from home is good, not because of a climatic change, but for the rest and change of scenery. We might not sleep well while traveling, but the temporary retreat from what we often consider a humdrum existence is refreshing.

Over the long haul, however, we are happier AND HEALTHIER right here at home. Most of us would suffer from the heat, if not strange diseases—strange to us, at least—if we went to the African veldt to live. And the African, in turn, would be miserable if he came here.

Weather doesn't affect everyone the same way. That fact was brought sharply home to us one fall when we held a church door open for an old man who had difficulty walking and used a cane to help him take slow, halting steps. The day was bright and clear.

"It's arthritis," he said. "I wish it would rain; then I'd feel better. The dry weather doesn't help me."

Countless thousands of others who suffer from the same affliction have more aches at the start of a rainy spell and seem to suffer less when it's hot and dry.

Many Northwest residents have gone south in search of relief from arthritis, yet a Los Angeles woman told me one New Year's Day (in Los Angeles):

"If you've got arthritis, stay away from here. The climate is bad for it." All of which only goes to prove that one man's dessert is another's poison.

I know of a father and son who came to Seattle from Minneapolis because the father was afflicted with hay fever each summer in Minnesota. The son, who had never had hay fever in the Midwest, got it as soon as he arrived in Western Washington, whereas the father's hay fever left him, never to return, within hours after he hung up his hat in his new home here.

An experiment at the University of Pennsylvania in a climate chamber (where temperature, humidity and air pressure could be controlled) proved that the effect of weather on arthritis isn't just another old wives' tale. A number of persons who had arthritis and claimed the ability to forecast the weather on the basis of their aches and pains became the guinea pigs.

The chamber—known as a Climatron—in which they lived for a time has walls ten feet thick, double plate-glass windows, and airlock "submarine" doors. The climate inside is completely under control.

At first the experimenters varied only temperature or rate of air movement, and none of those being tested felt any ill effects. But when the barometric pressure was reduced sharply (from 31.5 to 28.5 inches) and the humidity was increased from 25 to 80 percent, the majority of arthritics felt pain due to stiffening and swelling of the joints.

Seven out of eight rheumatism patients and three out of

four with osteoarthritis reported a worsening of symptoms. The pain abated after 25 hours of the high humidity and low pressure, indicating it was the change rather than the factors themselves which caused the condition. Lowering of the air pressure and an increase of humidity are conditions that often precede storms.

The weather has many other effects upon our health, many of which are little understood. The body seems to follow a cycle—daily or seasonal—in letting down or stiffening its guard against illness. It is known, for example, that certain diseases, including the staph infection, flare at specific times of the year, but no one knows exactly why.

Diseases that are transmitted from person to person by droplets (through sneezes and snorts), such as colds, the flu and other respiratory infections, are most common in winter and spring.

Polio and polio-like diseases reach their peak in the summer and early autumn, falling off when the rainy season comes. These are ailments transmitted by hand-to-hand contact, or the handling of things which have been contaminated by those who are ill.

As for skin diseases, some of them occur only in the summer—such as oak and ivy poisoning, severe sunburn, etc.—but certain others flare in the wintertime.

Rheumatic fever appears to reach a peak in April in the United States. Asthma patients seem to experience an increase in symptoms when the temperature takes a sudden drop, and the incidence of appendicitis throughout the nation increases where the temperature is rising but the barometer is dropping.

The dispersal of people in the summer has a lot to do with the decrease of illness. And their crowding together (as in schools) has much to do with the spread of diseases that flourish in the fall, winter, and spring.

The transitional days or weeks between winter and early spring are especially trying for many people who are ill. Another "bad" time of the year for many of us is that part of autumn immediately following Indian summer.

Declining temperatures and a shortening of the days

have a profound effect upon many animals and plants. Some experience suspended animation.

Do these same factors work upon humans? And how does one account for increases in illness after winter's back is broken? Is the body unable to adjust quickly to the climatic change?

The effect of electricity upon life is another one of the least understood things on earth, but scientists have made some interesting discoveries about human behavior and its relation to electrical activity in the atmosphere.

A report on the subject by the World Meterological Organization, an affiliate of the United Nations, indicates that both births and deaths rise during periods of high electrical activity in the air. So do traffic accidents.

All nerve cells in the body are electrically polarized, and there is a constant electric current flowing within the nerves and their fibers. But there appears to be a difference in the electrical flow within normal human beings and those who are mentally ill.

There seems to be a rise in mental illnesses during electrical storms. A study is now being made to see how mental patients react at times of intense electrical discharges in the atmosphere, and at times when solar flares are numerous.

Lightning is the most apparent form of electricity in the earth's air, but we don't need a thunderstorm to have electrical activity overhead. This form of energy exists at all times in air, water and our bodies. In fact, electricity seems to be inherent in life itself. It's no wonder we are affected by its surges.

Electricity has a definite correlation with the weather, and thus upon our health. The relationship between hot, muggy weather and heart failure has been demonstrated in the laboratory. And every doctor knows that an attack of angina pectoris can be brought on by walking on a cold day. It's not the walking so far as is known, but the chilly air that does the damage. The cold presumably produces general reflex vasoconstriction in some people, causing extra work for the heart. The reflex may originate in the receptors of the upper respiratory tract walls.

Lack of sufficient oxygen will cause damage to body tissues. Climbers of the world's highest mountains, such as Everest, must condition themselves slowly to the thin air. And above 18,000 feet a mountaineer will be in a zone where the body becomes weakened. Body tissues would deteriorate at elevations above 23,500 feet if one were to stay at such heights for a long period. That's why Jim Whittaker and others who climbed Mt. Everest made the trip from the highest base camp to the summit (elev. 29,028 feet) in the fastest time possible.

Back here, close to sea level, oxygen is in generous supply, yet sufferers from heart trouble, emphysema, and certain other ailments may need to breathe concentrated oxygen (always under medical supervision) from time to time.

Feeling "under the weather" is more than an idle saying. We know our health is affected by the weather, but we still don't know all the answers.

Seattle's Climate Ideal

Most health authorities agree that Seattle is a good place in which to be born and a great place in which to live.

Babies get a good start in life here, thanks to the city's pure water, clean milk, and comfortable climate. Most of them escape intestinal afflictions that plague infants in many other parts of the country.

Hot, humid weather is extremely trying on babies; so are extremes of climate. The youngsters get along much better here where sizzling heat and sultriness are rare.

The infant mortality rate for the entire United States was 15.1 per 1,000 births in 1976, the last year for which data are available. Seattle's infant death rate has consistently remained below the national level.

Health authorities say that the high rate of intelligence of people here is the main factor in keeping the death rate of infants low. The parents have a lot of savvy concerning infant care. But health officials also say (with pride) that this climate is mighty good for babies.

This climate is also mighty good for work (so there goes *one* excuse).

Most of us put forth our best physical efforts when the average temperature ranges between 60 and 65 degrees Fahrenheit. A typical day for this superior output would have a temperature at noon of 70 or a little more, and a nighttime low around 55 degrees. Add 70 to 55, divide by two, and you get an average, 62.5 degrees.

Early in this century, Ellsworth Huntington, a weather-conscious geographer and explorer, conducted tests among 500 factory workers in Connecticut and 3,000 in the South to see how the weather affected their work. He found that changes in barometric pressure had little effect on factory production. Humidity, he said, "possessed considerable importance," but temperature was the principal factor in altering productivity.

Huntington concluded that the ideal climatic condition embraced mild winters with some frosts, mild summers with the temperature rarely rising above 75 degrees, a succession of mild storms, and moderate changes of temperature from day to day.

All of us know from experience that a spell of extremely hot weather saps our strength, but if the heat lasts long enough—a week or two—we start getting used to it.

The first hot day doesn't greatly diminish our energy because the body is able to carry the impetus of previous cooler weather into the first part of a hot spell. However, the heat begins to "take hold" on the second or third day, and we perspire profusely. We also may begin to slack up a bit so far as physical effort is concerned.

All this happens because it takes a while for the human body to adjust to extreme heat or cold. Fortunately, our extremes of temperature in Western Oregon, Washington, and British Columbia don't last long. Isn't it great to have a flow of fresh ocean air break up a heat wave hereabouts?

Huntington was enthusiastic about the climate of Northern California, Oregon, Washington, and British Columbia. He placed us in one of the four best climatic zones in the world. The other three zones on his preferred list are (1) England, (2) New Zealand, and (3) part of South America, including southern Chile and portions of Patagonia, both in the South Temperate Zone. From what I've heard, parts of South Africa ought to receive favored status, too.

England and our northern Pacific Coast owe their climatic excellence to the fact that the ocean winds blow freely over the land. Huntington contended that a measure of storminess is a requisite in the weather in order to achieve stimulation, and I agree. But it's nice to get some summer calms, blue skies, and an occasional burst of heat, too.

Too much cold, as at the poles, and too much heat, as in the tropics, make it difficult for mankind to accomplish much. An American or Canadian from the North Temperate Zone could do little more than fight for survival in the dead of winter on the antarctic ice, and even then he'd have to have the aid of many products from home. Likewise, if we were suddenly transported to the equator, we would quickly wilt into a nonproductive state.

The best way to beat the heat locally (meaning the excessive warmth that comes so seldom) is to:

- Avoid an excess of alcoholic beverages.
- Drink cool water, but not too many iced drinks.
- Wear light, loose clothing. In other words, go tropical.
- Avoid over-exposure to the sun's direct rays.
- Keep a light, airy hat on your head if you work out-of-doors.
- And take a siesta, if you can, during the warmest part of the day.

Haven for Hay Fever Sufferers

August is hay fever time in North America—one of the worst months for approximately 10 million people in this country who suffer from allergic rhinitis, also called rose fever.

The allergy, which causes wheezes, sneezes, itching, tears, and other distressing symptoms of a cold, has nothing to do with either hay or roses, and there's no fever involved. But it's an illness and a scourge.

We who live in the Puget Sound area are lucky, though. Ragweed, one of the principal offenders, doesn't grow here. Neither does the Russian thistle, popularly called tumbleweed, which causes rhinitis in much of the West.

Nor do we have sagebrush, whose microspores may induce super-sensitive membranes to swell and weep.

Asthma is another affliction of humans often triggered or aggravated by the pollen of ragweed, tumbleweed, sage, and other plants that cast billions of dust-like particles into the air.

The Puget Sound region is a haven for many who suffer from hay fever in the Midwest, East and South.

Many of this area's permanent residents came here originally as visitors and were pleasantly surprised to find that their hay fever vanished soon after they crossed the Cascade Mountains, traveling west.

I don't mean to imply that hay fever is unknown here. We have many plants whose pollens cause distress, but the worst offenders don't like our climate.

A number of allergists, have found that warm, dry weather in the morning stimulates pollenization and thus increases the hazards. In the afternoon the hot, dry weather is less hazardous.

Medicines are available to alleviate suffering from pollen miseries. Climate control in homes and business offices helps, too. But the luckiest people are those who live in hay fever havens.

Climate control involves air conditioning, air ionization, and humidification.

Although hay fever strikes hardest in August, it may start in the spring when tree buds burst and continue until the first killing frost of fall. Those afflicted for months on end are those sensitive to many different pollens.

One ragweed plant can throw off as many as a billion grains of pollen, or enough to torture thousands of people. Any plant of the genus Ambrosia is known as ragweed in the United States. There are two principal types, the common and the giant.

Ragweed is found in this state in parts of Eastern Washington. The Russian thistle and sagebrush also thrive on the other side of the Cascades, but here on the west side, it is hay fever haven.

Puget Sound area residents *do* suffer from another seasonal ill, however—it's called Spring Fever.

It's a pleasant kind of "ailment" caused by the weather. The springtime warmth begins to do things to your body. You are likely to feel listless during spells of balmy weather, and full of pep when wind and rain are in blustery command.

The body suddenly finds itself compelled to adjust to increasing heat after having been conditioned to winter's chills. A change takes place in the blood vessels and the blood.

Grandma used to say that the blood gets thinner in the spring, and she was right, although the physiological changes are not that simple. Both the circulatory system and the blood manufacturing processes are undergoing transformation.

But it takes a while for the adjustment. Meanwhile, you may have, or get, spring fever. But don't worry. It isn't catching, it isn't dangerous. It's rather agreeable, although you might not get all the chores done.

Daydreams are a side effect. But they're nice, too, especially when the "dreams" involve hikes in the forest, fishing on your favorite trout stream, or lolling on the beach.

But let's face it. There's work to be done.

Smog and Temperature Inversions

Smog is a "dirty" word, or at least has a dirty meaning, but the reference is to atmospheric, not semantic, filth. All major industrial cities in the Far West are becoming aware of health problems caused by pollutants in the air.

Asthma and other upper respiratory infections flare when smoke, ash, hydrocarbons, and other substances collect in stagnant air. And virtually everyone, at one time or another, has suffered eye irritation from smog.

The word smog was coined way back in 1905, but not until mid-century did it receive widespread acceptance to denote a combination of natural fog and pollutants from the incomplete burning of fuel. Now dirty air alone, without fog, also has come to be known as smog.

Smog develops when an inversion exists in the atmosphere. An inversion is a lid of warm air over cooler air near the ground.

The industrial belts of Washington, Oregon, and British Columbia get occasional "dust caps," and they can occur in any season, even winter.

Fortunately the Pacific Northwest's rains do a good job of cleansing the air. And the numerous pushes of ocean air, hurried along by the wind, prevent buildups of atmospheric stagnation.

Inversions usually develop during times when the barometric pressure is high and the air is calm.

Los Angeles and her industrial neighbors in California were the first urban areas on the Pacific Coast to take drastic action against air pollution. They had to act because the Los Angeles basin, 60 miles long and 25 miles wide, was beginning to outgrow its air supply. Chimneys, automobiles, and incinerators were constantly hurling poisons into the air.

Topography and climate caused the Los Angeles dilemma. The basin is hemmed in on three sides by mountains. Cool air pushed in from the ocean, but hot desert air flowed westward over the mountains to create persistent inversions.

The inversions are still a problem there, but government and industry finally realized that something had to be done to prevent disaster. Factories began installing equipment to keep impurities from going up smokestacks. Trash burning was eliminated by law, and a widespread educational campaign was undertaken to obtain the cooperation of homeowners. The people cooperated by having their furnaces checked and (if need be) repaired to prevent poor combustion of fuels.

Strict antipollution laws were passed. Now, if necessary, industrial shutdowns can be ordered. Cities throughout the United States and Canada have taken similar action to curb pollution of the air as well as water. But smog is probably here to stay, off-and-on, when the winds die down and those pesky inversions develop.

Fog, in contrast to smog, is water distributed through the air in minute particles. The droplets are so small that they float instead of falling. It has been estimated that it might take as many as seven billion of the droplets to make a teaspoon of water.

Fog is a good example of a cloud at ground level. The droplets form when the water vapor in the air condenses on ultramicroscopic particles of sea salt, dust, or other nuclei. When the nights are long and the days short, as in winter, the fog that develops at night may last all day.

In summer, however, we often see only a shallow layer of ground fog, sometimes just a few feet above the earth, and it quickly vanishes under the influence of the warm sun. Ground fog of that type is a good sign that the day will be pleasant.

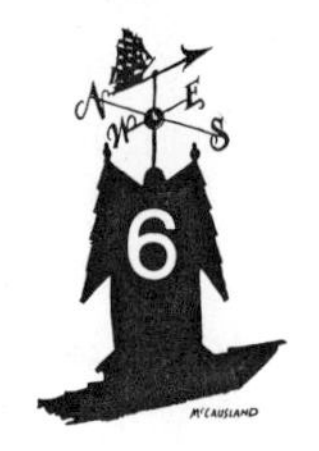

A Weather Eye on Sports

FAIR weather for baseball players was the topic of a column written for the *Seattle Post-Intelligencer* during the season of 1969 when the city's American League team was known as the Pilots. Games were played then in Sick's Stadium, where weather was a factor affecting pitching and hitting. Seattle's Kingdome is now the home of the Seattle Mariners, the city's second American League team. The Mariners' weather worries may be over, but every amateur league player still has to deal with the weather.

Baseball pitchers don't like to hurl on a sizzling hot day. That kind of weather saps their strength. But they also shudder at the thought of pitching on cold, wet days when the ball is hard to handle.

The average hurler would rather work on a cool, dry day than in the boiling sun. He feels stronger on a cool day (or night) and can do a better job with a fast ball.

However, the ball is likely to break better on a warm afternoon, especially if the humidity is high. The moisture in the air makes it easier to get some spin on the ball.

Those are comments by Sal Maglie, once a top major league hurler.

Norman A. Matson, Seattle's weatherman, and I spent part of an afternoon with Maglie one day. We wanted to learn how weather affects both pitchers and batters and under what conditions batted balls travel the farthest.

When Sal was pitching, he preferred a day when little or no wind was blowing and the temperature was around 75 degrees. If the wind was brisk, he wanted it to be moving from the third base side toward first. For him, it was a good day to throw his curve. He's a righthander.

In cold weather the ball is slicker and the fingers are less flexible, making it difficult to put spin on the ball, according to Maglie.

In 40-degree weather Sal used to feel much stronger than on a very warm day, and in such cool weather he'd go with his fast ball. He believes the average pitcher in baseball today also has more zip when it's cool, or at least not boiling hot.

Maglie recalls that he once pitched in 110-degree

weather in Veracruz, Mexico, and he quit at the end of seven innings, "thoroughly beat."

On another occasion he pitched in Philadelphia on a day when the mercury hit 100, and he lost 12 pounds.

The average pitcher will lose around seven pounds in a nine-inning game, Sal said.

Maglie is famed not only as one of the pitching greats in professional baseball, but as the guy who didn't shave on the day he was scheduled to pitch. Did he refrain from shaving to look mean?

"No, no," Sal said. "Of course my beard is dark and maybe I looked a bit rough, but my main concern was to avoid face burn. I wanted to devote all my thoughts to pitching and not be bothered with a tender skin."

Other baseball players (not Maglie) told me that Sal had a variety of deceptive pitches at his prime on the mound.

Sal acknowledged that he depended a lot on his curve but not at high elevations where the air is thin.

"The curve breaks better in Seattle than at Salt Lake City, Denver, Mexico City, and other high places," he said. "And when you get high enough, the curve isn't effective at all."

Sal recalled pitching a lot in Mexico City, where the elevation is over 7,500 feet and said he depended almost entirely on his fast ball there.

The thin air that's so hard on curve ball pitchers is good for batters. They can hit the ball farther at high elevations. The reason, according to the weatherman, is that the ball meets less resistance from the air.

Many people, including some baseball players, believe that a batter can knock a fast ball farther than any other pitch, provided he meets it with good wood. 'Tain't so, according to Maglie.

"The fast ball goes far if hit squarely, but the slider goes farther," Sal said. "The hanging curve virtually stops at the plate; it's ready for a long trip.

"A good batter might send a fast ball sailing 350 feet, but the slider (hit under the same circumstances) probably would go 400 feet."

Matson's calculations show that a ball hit at an angle of 30 degrees and an initial speed of 100 miles an hour would travel about 360 feet before the push of the air, plus gravity, would bring it down.

That deduction is based on weather data for a warm day, with a temperature of 95 degrees and air density of .071 pounds per cubic foot, an average figure. That would be considered light air.

Let's presume Don Mincher is at bat for the Pilots on this particular day. He takes a vicious cut at the ball and meets it squarely. The ball begins its ride at the 30 degree angle and a speed of 147 feet per second.

But the air and gravity start acting quickly on the sphere. In the first second the speed will be reduced to 120 feet per second, in the next second to 101 fps, in the third second to 88 fps, and in the fourth to 78 feet per second.

Gravity is working too, pulling the ball downward 16 feet in one second, 64 feet in two, 144 feet in three, and 192 feet in four.

The ball, headed for centerfield, arcs to the ground at 360 feet (a hit, we'll say), and Mincher legs it to third.

Now let's put Don to bat in a night game. The air is denser (0.77 pounds per cubic foot) and the temperature 50 degrees. There's a nip to the evening and Don, full of zip, takes a tremendous cut at the ball. Once again he lays good wood to it.

The ball sails off at the same angle, 30 degrees, and the same initial speed, 147 feet per second. But things are different. The ball meets greater resistance from the air. The momentum, drag, and other factors act faster. The only constant factor is gravity.

The ball's final speed in one second is cut from 147 to 117 feet per second.

Using the same laws of physics and mathematical formulas to trace the trajectory, Matson found that the ball would fall about 15 feet short of the daytime wallop. So the sphere drops for a hit 345 feet from home plate.

Don is speeding around the bases, but the centerfielder gets to the ball quickly and makes a good throw to the in-field. Where is Don? Well, maybe at second base. Why take a chance on being thrown out at third?

There you have it. One hit was a three-bagger, the other a double. Weather made the difference.

Seattle's weatherman delved deeply into physics to reach his conclusions. He also corresponded with Sir Graham Sutton, one of Great Britain's most distinguished meteorologists. Sir Graham, who has done research on the same subject, brought the matter to the attention of C.B. Dalsh of the physics department at the Royal Military College of Science in England.

The two Englishmen added more formulas in physics and mathematics, but Matson's are sufficient for me.

It isn't likely that any of the Pilots gave a hoot about formulas for the air's push on the ball. But every pitcher and all batters know that weather can make the baseball spin, sail, and fall in strange ways.

If you're an outdoor baseball fan you've heard the announcers say: "What a hit! It sailed over the fence at the 370-foot mark and was still rising! Imagine how much farther it would have gone in the lighter air down south!"

'Tain't so, Mabel. There's no lighter air down south, up north, back east or anywhere else, as a general rule, except up above. The higher you go, the thinner the air, but it's neither lighter nor thinner in different places on the Pacific Coast with any degree of regularity.

All of the parks in the Pacific Coast Baseball League are fairly close to sea level, so the air at San Diego, Los Angeles, Hollywood, San Francisco, or Sacramento isn't likely to give the long-ball hitters a special break.

If the weather scales are tilted at all in favor of the hitters, the advantage MIGHT exist in the Northwest. The air in Portland, Seattle, and Vancouver, British Columbia, may be lighter more often than the air in California because it is often moist.

If you have two different air masses, one moist and the other dry, the former will be the lighter, provided temperatures are the same in each. There are other factors that affect the air's density and weight, but as a rule moist air is light air.

Barometric pressure is high (the air presses down upon us more heavily) when it's sunny. The pressure is lower during stormy weather.

There are times when Seattle's air is extremely dry during the baseball season. On those occasions the pressure would be high and the air would be dense. But California has the same kind of dryness. So you can't pinpoint the air in any one ball park as permanently light or heavy unless you are comparing air at vastly different elevations.

Denver, perched a mile above the sea, and Salt Lake City, at an elevation of more than 4,200 feet, are good examples of places where the air is light. Long-ball hitters often had field days in the rarefied air.

Normal sea level pressure is close to 30.00 inches. The pressure at 900 feet is approximately 1/30th less, or 29.00 inches. If you rise 900 feet higher, the pressure lowers to 1/30th of 29, or 28.03. The decrease continues at the same approximate ratio for each 900 feet of altitude.

This formula gives a pressure of about 25.62 inches for Salt Lake City and around 24.68 for Denver. It's no wonder that baseballs sail through the air with the greatest of ease in those cities.

If you or I make a sudden change from sea level to a great height in the mountains, we're likely to get sleepy, giddy, or ill. We may gasp for air and tire quickly. Our bodies are telling us, under such circumstances, that we're not physically equipped to make quick climbs out of the depths of the atmosphere.

Medical science has found that residents of places like Cerro de Pasco, Peru, at 14,385 feet, have bigger chests, thicker hearts, larger blood vessels, and thicker blood than lowlanders. There are numerous other differences, too, in their bodies. Nature has built them for life in the thin air.

A native of Cerro de Pasco would be uncomfortable at sea level. He'd feel a confining weight. His whole body would cry out in protest. He'd soon want to go back to the heights. He'd be as much out of his element at sea level as a Seattleite in Cerro.

We're wondering if they play baseball at Cerro. It ought to be a haven for weak hitters. Every bunt would be a home run.

Fair Weather for Fishermen

If you're a fisherman, you should be weather-conscious. The fish are.

Factors involved include temperature of the air, temperature of the water, wind, sky conditions, and state of the river. The stream can be too murky, too high, too low or (in winter) too cold.

Knowing how closely a fisherman's success or failure is tied in with the weather, I asked William J. Allyn, who is both a meteorologist and an angler, for his views on fishing in relation to the weather.

Let's talk about the wily steelhead for which the Pacific Northwest is famous.

Allyn, formerly a public service forecaster for the weather service, was among the first to suggest that weather data from Stampede Pass (elev. 3,958 feet in the Cascade Mountains) be included in The *P-I's* daily weather table "for the benefit of both fishermen and skiers."

It's obvious that the temperature and type of precipitation, if any, would be a clue to skiing conditions elsewhere in the mountains. But how does such information benefit anglers?

"Ah," said Allyn, "it's mighty important. Heavy rains and above-freezing temperatures mean that lowland rivers might rise, and become roily, whereas cold weather in the mountains would indicate that the rivers probably would become clear and low, or at least lower than when the runoff is heavy."

First in importance for winter fishing is the clothing you wear. Allyn suggests thermal underwear, warm socks, insulated boots, a cap with earflaps, a woolen shirt, and gloves.

I asked him if a woolen face covering, with slits for eyes, nose and mouth, was a necessity and he said: "When it's so cold you have to wear such a contraption, it's too cold to go fishing."

Murky water is bad for steelheading, according to Allyn, because "The fish has to see the lure before he can strike it."

The water can be too clear, though. Such is often the case after a long stretch of cold days. Allyn knows that the fish can see you thrashing around in the water, and he suspects they can see you approaching the river, too.

"A heavy-footed, fast-walking angler can spook the steelhead long before he walks into the stream," Allyn said. "So make a quiet approach to the river."

Wind is a problem, especially if it's blowing upstream toward you. When such is the case, Allyn looks for a place with less exposure to the wind, perhaps a section where trees or a bluff give protection.

Water temperature is a major factor in steelhead fishing, according to fisheries scientists.

It has been noted that fish are lethargic when the temperature is in the 30s. At and above 42 degrees they are more active, and angling is better.

This is not to say that steelhead don't move around below 42 degrees. Many a migration upstream from the ocean takes place when temperatures are in the 30s. But the movement is greater at and above 42 degrees.

It has been noted, too, that steelhead move in greater numbers over the fish counting device at Bonneville Dam on the Columbia River when the water temperature is 42 or more.

A representative of the Game Department said temperatures in the middle or high 40s are better than lower temperatures for steelhead movements. He agreed that temperatures in the 30s slow up the fish. "Fish are inclined to hole-up during extremely cold weather," he said.

Steelhead may be on the move during flood stages of our rivers, but peak runs don't occur in such weather.

"Changes in water flow make a big difference to the fish. When the water is very low and clear, they wait for a rise in the river. The water level seems to be even more important than the temperature at times."

Warm-weather fishing is something else again. "When temperatures rise into the 70s, fishing falls off, especially in the lakes. Fish once found near the surface and in the shallows disappear, seeking colder water. But in the fall when it's cool they return to the shoals in search of food."

The man who taught Allyn how to catch hardheads, offers this advice to steelheaders:

1. Keep your hooks sharp.

2. Don't use too long a leader. Steelhead rest and travel near the bottom, and a long leader will put the bait too far above the fish, even if your lead is bumping bottom.

3. Set the hook if you have the least suspicion that a fish is nibbling at your bait.

4. Don't smell too good. Fish don't like the human odor. Rub some old stinky fish eggs on your hands so you don't pass on human odor to your lure.

5. Use good equipment, and keep it in first-class shape. You never know when you're going to hook a big, wild steelhead.

6. Don't be selfish. Share your knowledge, and others may do the same for you.

7. Don't waste a lot of time trying to retrieve badly snagged gear. It's better to break off, retie, and get back in business. Moreover, it is senseless to risk your life fighting fast water in order to save a 19-cent lure.

8. Dress comfortably and properly for the cold.

9. If the rivers are out of shape, go home early and spend some time getting your tackle in the best possible shape. You'll be building points for future trips when conditions are better.

Now hear this bit of philosophy from the pen of Izaak Walton:

"Angling may be said to be so like the mathematics that it can never be fully learnt.

"As no man is born an artist, so no man is born an angler."

Forecasting

Proverbs

SOMEONE once said that every weather proverb bearing the whiskers of age holds a grain of truth. That's a reasonable conclusion if we limit the efficacy of each bit of weather lore to its source region or areas with comparable topography.

My research indicates that most Old World proverbs are not exportable to our Pacific Northwest, although many will apply in a general way. Some meteorological sayings, for example, refer to the dandelion, scarlet pimpernel, and other flowers "forecasting" wet weather, and those observations are correct because increasing moisture in the air, here or anywhere else, will cause those plants (and certain others) to close their flowers. Some of the rainbow proverbs also apply in many places here and abroad. But the world's weather lore contains many contradictions.

For instance, Chapter 27 of the Book of Acts tells of a shipwreck "at a place where two seas met" off the island of Melita while Saint Paul was on his way to Rome. The south wind had blown softly, according to the account, but "not long after" there arose against it a tempestuous wind called Euroclydon. Euroclydon was a name given in those days to an east-northeast wind in accordance with the custom of personifying the winds.

The ship ran aground close to the mouth of a creek, and the forepart "stuck fast and remained unmovable, but the hinder part was broken with the violence of the waves." Those who could swim cast themselves into the sea and got to land. The others used boards or broken pieces of the ship to carry them ashore.

Paul and his companions reached Italy months later on another ship.

I haven't yet found a Biblical reference to a kind or gentle east wind. Psalm 48:7 says: "Thou breakest the ships of Tarshish with an east wind," and Ezekiel 27:26 declares: " . . . The east wind hath broken thee in the midst of the seas."

A hot, dry east wind also figured in one of the dreams of Pharaoh, the king, who made Joseph ruler under him

"over all the land of Egypt." Pharaoh "saw" seven ears of corn come up on one stalk, rank and good, followed by the growth of seven thin ears which were "blasted with the east wind." Joseph interpreted that dream as a warning that seven bad years (drought, poor crops, and hunger) would follow seven years of plenty.

No doubt an east wind was then, and still is, a searing unwelcome wind in much of the Middle East. Our east winds in Western Washington, however, are hot in summer, cold in winter. It's our southwest winds that bring wild weather to the Pacific Northwest.

My friend, the Rev. Dr. Cecil Ristow, retired United Methodist pastor (formerly at University Temple in Seattle), points out that rain, snow, lightning, wind, and rainbows are mentioned many times in the Bible.

Psalm 135:6-7 tells us that: "Whatsoever the Lord pleased, that He did in heaven and in earth, in the seas, and all deep places.

"He causeth the vapors to ascend from the ends of the earth; He maketh lightnings for the rain; He bringeth the winds out of His treasuries."

The writer of Job says in Job 37:9: "Out of the south cometh the whirlwind: and cold out of the north." How true that is—all of the northern hemisphere's cold weather comes out of the north, even though it may edge in sideways.

One of the oldest weather proverbs concerns a red sky. We have it in many forms, including this one:

Red sky at night, sailor's delight;
Red sky in the morning, sailor take warning!

That one goes back to the Bible, too. In Matthew 16:2, 3 we find mention of it in Jesus' answer to the Pharisees and the Sadducees, who asked that He show them a sign from heaven. He answered:

When it is evening, ye say, it will be fair weather: for the sky is red.
And in the morning it will be foul weather to-day: for the sky is red and lowering . . .

One of the proverbs I like best, and find fairly reliable, concerns a change in the cloudiness when many clouds are milling around in the sky on a blustery morning. The adage tells us that the weather will clear up if a patch of blue, big enough to make a Dutchman's trousers, suddenly appears. The reference is to the baggy pants of old Holland, so you'd have to spot a sizable hole, showing the blue vault above the clouds, for general clearing to occur. That proverb, though not infallible, proves correct more than half the time.

I find no merit in the adage that "many hips and haws" indicate a severe winter coming. Hips are the fruit of the rose, and haws are hawthorn berries. The weather of the preceding spring and summer helps determine how many hips and haws will develop, just as the right kind of weather during the growing season is a major factor in the abundance of nuts, fruits, and other crops.

Many proverbs refer to rain. Most are meaningless. I like this saying, "When God wills, it rains with any wind." And the following jingle is worth mentioning:

The rain it raineth every day
Upon the just and unjust;
But mostly on the just because
The unjust hath the just's umbrella.

Some insects, including ants, are said to be good weather prognosticators. However, I'm inclined to think they judge the times to take refuge and the times to go about their daily chores by the moisture content of the air. They may sense a coming storm by an increase in the humidity and go into hiding. I doubt that any other prescience enlightens them.

When rain is due, spiders have been seen tightening the threads of their webs. And they often lengthen those supports if good weather is at hand. Here, again, changes in the humidity probably spur them to action.

If you've ever seen a woolly bear caterpillar, you may recall that it has three distinct color rings. Each ring has hair of a different color.

A belief persists that each segment of the woolly bear's

"coat" represents a third of the coming winter, and the darker the color, the colder and nastier the weather will be.

My informant for one year said her woolly bear forecast a mild November, but a cold, snowy December and January. The nearly black hair indicated that chilly weather could be expected from "about December 20 to about January 20," she told me.

That woolly bear's "forecast" was questionable. A schoolboy who collected insects "for the fun of it" showed me two other woolly bears in a bottle. One had a ring as black as a new moon; the other displayed three shades of brown hair, fairly light.

Which woolly bear to believe?!

The best sign of imminent good weather is the presence of dew glistening on the grass before sunrise on a spring or summer day. If you see that dew it's a safe bet that you can plan an outdoor picnic.

Halos around the sun or moon usually mean that the high clouds, consisting of ice crystals, are increasing and probably lowering too. And the increase in cloudiness often means a change for the worse in the weather.

Scattered cumulus clouds (the cottony kind) on a summer day are known as, "clouds of fair weather." But if conditions in the air are unstable, cumulus clouds may develop into thunderheads and, ultimately, into cumulonimbus clouds, the big, ugly-looking monsters that produce thunder, lightning, hail, and heavy rain. So, remember:

When clouds appear like rocks and towers,
The earth's refreshed by frequent showers.

The moon, that friendly orb of mystery in the sky, has been the theme of songs and adages ever since man first looked upon lovely Luna shining upon the earth. The Scots have a saying that rain or snow can be looked for if the moon has a white look or when her outline isn't clear. And the Zuni Indians of our country offer these lines:

The moon, her face if red be,
Of water speaks she.

I've heard people say that the weather tends to be clear when the moon is full. They're wrong. The moon is full at the same time everywhere, but the weather may be dry in one place, soaking wet someplace else, and cloudy in the next county.

If we see the full moon, the weather obviously is fair for the moment. But many a full moon comes and goes in the Pacific Northwest without anyone getting a glimpse of it because of inclement weather. I like this rhyme:

The moon and the weather
May change together,
But change of the moon
Does not change the weather;
For if we had no moon at all,
And that may seem strange,
We still would have weather
That's subject to change.

All of us have heard, and some believe, that nature prepares animals for harsh winter weather by providing an abundance of fur or hair in the fall. Every animal is born with a specific number of hair follicles, and there's no way the number can increase.

Climate is a factor, though, in preparing all of us, including our pets and all the wild creatures, for average weather conditions. People born and reared in the Pacific Northwest get acclimated for our average weather but might have a difficult time adjusting to conditions farther north, farther south, or even far inland.

You and I might not survive on an arctic ice floe without a lot of artificial heat to ward off the chill, but the polar bears, the snowy owls, and the Eskimos would make out all right.

I can't subscribe to such sayings as, "A warm January, a cold May," and "A cold January, a feverish February, a dusty March, a weeping April, and a windy May, presage a good year and gay." But I'll go along with these European sayings:

January warm, the Lord have mercy.
January blossoms fill no man's cellar.
A January spring is worth nothing.

Those proverbs apply because premature growth due to warmth in January is liable to suffer from frosts and chills in succeeding months.

Our English literature is rich with observations about climate and weather, principally because weather always was, and probably always will be, a major adversary of the British, Canadians, and Americans. Weather lore is abundant, too, in the literature of the Scandinavians, Germans, Hollanders, and other peoples of northern Europe.

But most of the great writers since Chaucer's time confined their meteorological comments to brilliant descriptions of seasonal weather or lilting references to nature's pulsing rhythms.

Whittier wrote, "Night is mother of the day, the winter of the spring." And Ella Higginson penned these lines:

Oh, every year hath its winter,
And every year hath its rain—
But a day is always coming
When the birds go north again.

Edward Estlin Cummings remarked that, "The sweet, small, clumsy feet of April came into the ragged meadow of my soul." Sir Henry Newbolt said, "April's anger is swift to fall, April's wonder is worth it all," and William Browne observed, "There is no season such delight can bring as summer, autumn, winter, and the spring."

I salute all weather-watchers. They're my "blood brothers." I urge all of them to delve deeply into the weather lore found in proverbs and in our literary inheritances. But don't expect the proverbs to be exact "forecasters."

Almanacs

How good is a weather forecast in the average almanac or on one of those calendars that offers day-by-day forecasts?

The answer, according to Marvin Magnuson, former area climatologist for the National Weather Service, is: "About as good as a random guess, provided the guesser uses climatological data to buttress his prediction."

Magnuson isn't opposed to almanacs per se. He finds them full of interesting information such as recipes, postal regulations, fishing laws, dates of eclipses, jokes, puzzles, etcetera. But he shudders at the thought of anyone giving serious consideration to an almanac prophecy that goes beyond climatological normals.

Magnuson concedes that a forecast about the weather for a particular day next month, or a thousand years from now, could prove correct. And he acknowledges that a prediction of rain—or even hail or snow—on a summer day might turn out to be right, too.

His main concern is that some almanac readers start wondering, or even worrying, when they see a startling prediction like, "The storm of the winter—20 inches of snow—will occur on (a specific day)."

Most people wouldn't take much stock in that kind of long-range forecast, but Magnuson heard enough about such a prediction in the mid-1960s to cease smiling at the thought of it. The forecast, made months ahead of time, called for 20 inches of snow on one particular day in January. And a big snow hit Seattle on schedule. The so-called "accuracy" was identical to what one might achieve by intelligent guessing, according to Magnuson.

Robb Sagendorph, former president of Yankee, Inc., publishers of the *Old Farmer's Almanac,* told us that his almanac had fairly good luck in hitting "extremes of weather." He said the forecaster, Abe Weatherwise, uses an inductive rather than a deductive approach. He pointed out that the forecasts are made "more than a year ahead of time," hence: "We cannot recommend total faith in this forecast, and don't mind at all how much anyone jibes at it."

Yankee, Inc., is still a thriving business, and its almanac is a weather "bible" for people throughout the country. I don't take much interest in the forecasts, though I read and enjoy them. Best of all, however, it offers a full measure of good humor and puzzles to while away the winter nights, and you can *totally* enjoy that!

Nature Tells All

Science isn't likely to receive any tips from Nature on what the weather will be like in any season, despite the sensitivity of many plants, animals and insects to changes in humidity, pressure, and temperature.

Nor is it reasonable to expect that summer's crops, whether meager or plentiful, will provide clues to the next winter's weather.

It is true that many plants close their flowers before a rainstorm. The world's weather lore is full of such sayings as: "The daisy shuts its eyes before rain," Benjamin Franklin.

Flowers that close just ahead of a shower are sensitive to heavy concentrations of moisture in the air. It has been determined that the pimpernel begins to close when the humidity reaches 80 percent, and the chickweed starts protecting itself by curling its leaves when the air becomes 82 percent saturated.

You and I, looking at the sky and checking the wind's direction, probably can tell more quickly than the pimpernel whether rain is on the way.

Two plants, the common purple lilac and the honeysuckle, have been used in the West (along with the thermometer) to help climatologists, gardeners, and farmers determine whether spring is early, late, or "on time."

Lilacs begin blooming in Seattle late in April and the floral display continues into May. If the flowers appear in early April, the season is early.

Lilacs are in blossom somewhere in the West from early March through June. They bloom earliest (around March 11) in the extreme southwest, many miles inland, and latest at 7,500 feet in the mountains.

The lilac was first choice for the phenological survey because it is so widespread in the West and is day-neutral, meaning that the day's length has little effect upon growth, development of the flowers and the onset of dormancy. The time of flowering is largely dependent upon the temperature.

Lilacs aren't grown in some parts of southwestern Arizona and southeastern California, however, because of inadequate chilling during winter. The lilac, like most of our deciduous shrubs and trees in the Northwest, must experience enough cool weather to give it a non-growing rest period.

Cats, dogs, horses, cows and many other animals, are affected by the *air pressure* (revealed by the rising or falling of the barometer), and by the humidity. Grazing sheep and cows often crowd together just before a rain. Horses will stretch their necks and sniff the air prior to a storm. Animals in a barn often become restless when foul weather is approaching.

Mountain goats are said to graze downward just before a storm and upward if the weather is going to remain fair. Old goats and sheep eat greedily before a storm.

But such activities are observed only a short while before a change in the weather. They aren't long-range "forecasts."

An English proverb says: "When a cat sneezes it is a sign of rain."

And a Spanish legend tells us that: "Sailors don't like to see the cat on board ship unusually playful or quarrelsome because it means the cat has a gale of wind in her tail."

The weather lore of Greece declares that "if a cat licks herself with her face turned north, the wind will soon blow from that dangerous quarter."

The word "dangerous," in that proverb indicates that Greece's worst weather comes out of the north.

Among Nature proverbs that have merit are these: "The silver maple shows the lining of its leaf before a storm (because the wind rustles the leaves)." "When seagulls take shelter inland, the sea is stormy."

The crickets also have a clue to contribute. They will sing the temperature for you if you don't have a thermometer. Here's the way to take the "reading:"

Count the number of chirps per minute, divide by four, then add 39. If, for example, you noted 160 chirps in 60 seconds, you would use one-fourth of that figure, or 40, and add 39 to give a "reading" in degrees Fahrenheit of 79.

This formula dates from olden times. It is said to be most reliable when applied to the high-pitched tremulous chirps of the snowy tree cricket which is found in Eastern Washington but not on the west side of the Cascades. However, the song of any cricket will do in a pinch.

The weather affects our insect friends and enemies a lot more than it does us, as a rule. A warm season can cause tremendous increases in certain pests (like the mosquitoes), whereas cold weather might wipe out numerous generations of insects.

Insects are cold-blooded, so they respond directly to temperature changes. That's why the "songs" of various insects can be translated directly into temperature "readings."

Temperature may combine with moisture to limit the numbers of insects or their habitat. Most tropical bugs will die at once in freezing temperatures; yet many northern insects can stand extreme cold.

In some respects, insects are more adaptable to changes in oxygen supply and pressure than humans. Mosquitoes have been found alive a mile in the air, and aphids at 7,000 feet.

Several other bugs were collected in special traps attached to airplanes flying at higher altitudes. They were dead when removed from the traps, but some may have been alive before their capture. This collection included thrips at 10,000 feet, and the striped cucumber beetle and a gnat at 11,000 feet.

Those are just a few of the insects found high in the atmosphere. Most of the high-flyers are taken for a ride in sharp updrafts. Even wingless species, including bristletails (silverfish) and springtails have been found at 8,000 feet or higher.

The wind, which helps to scatter insects, also reduces the numbers of some bugs by blowing the insects or their eggs off the host plants. This often happens to the corn earworm and the corn borer.

West of the Cascades we can thank our weather for the

lack of many bugs that plague other parts of the nation. We don't get the intense heat that many insects need to thrive.

Do Your Own "Five-Senses Forecasting"

Our eyes, ears and nose are prepared to tell us a lot about the weather. Unfortunately, these perceptive senses are dull in many of us city folk.

People who live close to nature—the farmer, the forester, the fisherman, and many others—often become proficient in "smelling" weather changes and in translating some of the sounds of wind and water into tolerably good forecasts.

Many a rural resident can tell you, without looking at weather instruments, that the air pressure has fallen and rain is on the way. His "barometer" may be a stagnant pond, a peat bog, the ditch alongside the county road, or a pile of decaying matter on the farm. An increase in the aromas from one or more of these barometers indicates that the earth is "exhaling" more vigorously as a result of lessening air pressure.

Heavy air keeps many smells close to the ground, whereas light air allows them to rise higher into the atmosphere. You'll know what we mean if you've ever walked through the forest just before or during a rain and caught the sweet smell exuding from devil's club leaves and stems, from the fallen needles of the pines, the firs and the hemlocks, and from the trees themselves.

The barometer is usually low just before and during a storm.

Forests often "murmur" before a rain, and some mountains are said to rumble or "talk." The murmuring of the forest is the collective sounds of leaves and branches "singing" as the wind blows. An increase in the murmuring means the wind is picking up. The tones that come from thin fir needles are high-pitched, and those from bulkier needles are low in tone. A low, muffled sound emanates from orchards and groves of trees whose leaves are broad, like the maples.

The mountains that "talk" are merely relaying the

sounds of the wind as it cuts through canyons, swirls around peaks, or pushes into crevasses. Many mountains do their "talking" in advance of storms, thus giving nearby residents warning of a weather change.

One of the weather proverbs warns us:

When the forest murmurs and the
mountain roars,
Then close your windows and shut your doors.

When the wind is increasing at sea and causing the waves to mount in size, the sound is often heard many miles inland. The people who hear the distant ocean "raising a fuss" can often tell that a storm is coming, and they will know many hours ahead of the storm's arrival.

Distant sounds usually are heard with unusual clarity before the weather changes for the worse.

Train whistles that aren't heard at all in clear weather may sound as if they're up the road a piece just before a rainstorm. The sound travels better in the moist air, so there's good sense to the maxim: "A good hearing day is a sign of wet."

Here are some other signs that will tell the alert weather watcher what kind of day to expect:

Ground Fog—A low, thin layer of fog that hugs the ground is an indication that the day's weather will be good. Ground fog develops in the early morning. It is seldom more than four or five feet deep, often less, and disperses quickly after sunrise.

Dew on the Grass—This is another sign, like ground fog, of fine weather. Both form when skies are clear. The cooling of a thin layer of air next to the earth causes the moisture to condense.

Sun and Moon Halos—A ring around the sun or moon is formed by the refraction (bending) of light rays as they pass through high, thin clouds made up of ice crystals. Such clouds usually are thickening and dropping at the time halos are seen. The thickening of the clouds ordinarily means a change for the worse, maybe rain.

Visibility—Distant shores seem nearer before a rainstorm. The haze that forms during a spell of clear weather is dispelled by faster moving, cleaner air.

Clear Overhead—If the moon is in sharp outline and the stars are numerous and bright, expect a good day on the morrow. But if the moon and stars are dull, chances are that unsettled or rainy weather is around the corner.

Flying Birds—Hunters know that waterfowl fly higher in good weather than in bad. A stormy day is a good day to hunt ducks and geese. Close to Puget Sound, we often see gulls circling high overhead just before a storm; they're enjoying themselves, "riding" the wind, but they will head for cover when the storm comes.

Here in Washington, a cloud cap over Mt. Rainier is nearly always a sign of rain, probably within 24 to 36 hours. This type of a cloud, which seems to be stationary, is actually forming on the windward side and dissipating in the dryer air to the north. It means that a south or southwest flow of moist air is headed our way.

Storm Centers—If you want to know in which direction a storm center lies, stand with your back to the wind and stretch out your left hand. The direction in which you are pointing is the approximate direction of the "blow." That rule applies everywhere in the northern hemisphere.

If you also want to determine the approximate speed of the wind without a measuring device, you can use a formula worked out by rangers in the United States Forest Service:

Wind less than 1 mile an hour—Smoke rises vertically; no movement of leaves on trees or bushes.

Wind 1 to 3 miles an hour—Twigs of trees move gently; small branches of bushes sway; weeds and tall grasses swing and sometimes bend; wind vane barely moves.

Wind 4 to 7 miles an hour—Trees of pole size (in the open) sway gently; you feel the wind on your face; loose scraps of paper move; small flags and pennants flutter.

Wind 8 to 12 miles an hour—Trees of pole size (in the open) sway noticeably; large branches of pole-size trees (in the open) toss about; tops of trees in dense stands sway; the wind extends small flags; a few crested waves develop on lakes.

Wind 13 to 18 miles an hour—Trees of pole size (in the open) sway violently; trees in dense stands sway noticeably; dust is raised from roads.

Wind 19 to 24 miles an hour—Branchlets may break off trees, especially if they are brittle; inconvenience is felt in walking against the wind.

Wind 25 to 28 miles an hour—Your progress is impeded when walking into the wind; loose shingles may be blown off roofs; all small and medium-sized trees are violently agitated; some trees may be severely damaged.

The scale ends there, but virtually every part of the United States has seen damage caused by winds of stronger velocity. Gales, which pack winds from 39 to 54 miles an hour, often uproot trees and blow down utility lines.

Winds from 55 to 63 miles an hour are called "storms," and winds of 74 or more miles an hour are described as "of hurricane force." A true hurricane is a particular type of tropical storm with a gigantic system of whirling winds.

Weather Instruments for the Do-It-Yourself Forecaster

A barometer, a Six's-type thermometer, or even an antique weather vane can add a good deal to your weather-watching enjoyment (and they won't hurt the accuracy of your forecasts any, either).

A barometer acts as a weather "prophet" by registering the air pressure as it rises and falls above us. Slight changes usually make no appreciable difference, but sharp decreases or increases in the pressure are significant.

The instructions that come with your barometer may tell you to adjust the instrument for altitude. You can forget that adjustment. The Weather Service converts the barometer pressure to sea-level readings, therefore you should have a sea-level reading.

The barometer is easily set by adjusting a screw at the back. If, for example, the pressure is 29.92 inches, move the screw until the main hand lies slightly past 29.9. You will find the face of the instrument marked off in inches and tenths of inches, so the hundredths will have to be determined by the eye.

Another hand (often red) is adjusted by moving a knob in the middle of the barometer's face. This hand, which won't move unless you move it, is used to keep a record of the barometric changes.

After setting the principal pointer to the correct pressure, twist the front knob until the other hand lies over it. Then, later in the day, check to see how much the main indicator has risen or fallen.

The words "fair," "change," and "rain," or similar designations may be on the face of the instrument. Sometimes the weather may change as indicated; other times it will be raining when the barometer is high, and dry when the pressure is low. It's interesting, nevertheless, to watch the ups and downs of the instrument.

Remember this: sharp increases or decreases in pressure usually are indicative of weather changes. The little changes don't mean much as a rule.

We can't expect the barometer to do any more than measure the pressure directly over us, but it may offer a clue to the weather coming tomorrow.

If you don't have anything better than an 89-cent thermometer, you're missing another one of the pleasures of weather-watching.

How warm was it yesterday? Is it likely to freeze tonight?

Those questions are asked almost daily in the average American home, yet no one in the family is likely to have the temperature data for your immediate neighborhood.

Sure, the National Weather Service report is available in a conspicuous spot in your newspaper, but those figures are for the general area. How about the data for your own yard?

Don't misunderstand, I'm not opposed to 89-cent thermometers. Many of them are fairly accurate. But they're rarely hung properly. And they don't give the readings that mean the most.

It hurts me as much to see a home thermometer tacked onto an outside wall as it does to see a candy wrapper on a mountain trail. Each is out of place.

The thermometer I recommend for household use (outside) is known as the Six's type, named for James Six who invented it about 1782 in Canterbury, England. It's a maximum-minimum thermometer with two sets of figures and the recording element in the U-shaped tube.

The figures and an indicator on the left record the low temperature and those on the right the high. The minimum temperature of the day, week, month, or whatever period you're checking will remain in the tube until you reset the instrument with a magnet. The same goes for the maximum reading.

With an ordinary thermometer you have to be right in front of it at the moment it reaches the day's maximum or minimum, and that's inconvenient if you have other things to do.

Not so with the Six's. If the low occurs at 4 a.m. while you're asleep, the reading will remain for you to see at your convenience.

If it's the maximum you want, you can go to work, the corner store, or London, England, and know that the reading will remain recorded in the tube until you reset the indicators.

The Weather Service uses more sophisticated instruments to record temperatures. The maximum thermometer is one instrument and the minimum another. Each requires a special support and both are kept in an instrument shelter that keeps the sunshine out but allows air to circulate freely around the instruments. The whole setup is costly and, for home use, unnecessary.

The Six's will do nicely, giving the weather fan enjoyment beyond measure. Warning: don't hang a thermometer on a wall or in an area where heat from the house will alter the readings.

You can put your thermometer on a bracket extending down from one of the eaves of your house, or in a rose trellis. Or hang it under a tree. Above all, don't let the sunshine fall upon it.

(By the way, have you noticed that the forecasters now give specific figures for the anticipated highs and lows of temperature? No longer do they say "low 45-50." Instead the forecast may read, in part, like this: "low about 45, except in suburbs where it may be 10 degrees colder.")

A maximum-minimum thermometer is a versatile instrument. It will warn you when sensitive plants need mulching or other protection from severe weather. It will tell you when you had better insulate exposed outdoor waterpipes to guard against freezing.

And it can serve as a guide to help you determine how your backyard temperatures differ from those at Weather Service stations.

The Six's thermometer, if carefully studied over a period of time, will give you some pretty good clues to your own microclimate. You'll have to make comparisons under various weather conditions, of course. Rainy days might produce temperatures that are virtually the same in the same area, whereas clear days (and nights) could produce startling differences between different areas, or even between hilltops and valleys.

If you're a weather enthusiast, you'll find a Six's thermometer one of the best investments you ever made. Several companies make them, although some dealers are mystified if you use the name Six's. Ask for one instrument that gives both maximum and minimum readings.

You'll enjoy it.

Weather Vanes

Even older than Six's thermometer, the weather vane, a useful and picturesque instrument, has come close to extinction in American architecture, but it might be making a comeback. There appears to be a revival of interest in mounting "weathercocks" on homes, barns, and buildings, undoubtedly sparked by America's bicentennial celebration in 1976.

The bicentennial helped by focusing attention on ornaments and furnishings popular when our nation was young, so maybe the "weathercock" of olden times can eventually be taken off the endangered list. Some of the ancient wind vanes are among the most treasured antiques in America.

However, you still have to scout around to find weather vanes of any age, new or old, in residential areas. They are virtually nonexistent on new public buildings. For instance, I don't recall seeing any high-flying, easily visible, wind vane in downtown Seattle until Ivar Haglund, new owner of the Smith Tower, had a huge one, shaped like a fish, installed above the topmost pinnacle of that 42-story structure.

Our Pacific Northwest states and the Province of British Columbia still have a few vanes dating from pioneer days. You also will find, by diligent searching, some ultramodern vanes conceived by imaginative hobbyists.

I've come across vanes shaped in the forms of pheasants, flying ducks, horses, cowboys, and full-rigged sailing ships. Most of them are in older neighborhoods where affluent families live.

The British, who are more inclined than we are to savor and preserve the past, probably have ten times as many wind vanes as there are in all of our 50 states. And some of those vanes are among the world's finest examples of this ancient art form. France, Germany, Belgium, the Netherlands, and other mainland countries in Europe also have numerous old-time "weathercocks" still in service.

The "weathercock," symbolized by a crowing rooster, is an ornament dating from the ninth century when Pope Nicholas I suggested that the top of every cathedral, abbey, and parish church in Christendom bear the likeness of a rooster. He made the proposal to remind mankind that Jesus' beloved apostle, Peter, displayed human frailty just before the crucifixion when he thrice denied knowing Christ, being fearful that he, too, might be seized and slain.

The fish came into use as a wind indicator about the same time as the rooster because it was the chief symbol of the early church.

Weather-watching is an interesting hobby, and it is more fun if you have a "weathercock," or one of the modern offshoots, to reveal where the wind is going and from whence it came, especially on those occasions when it seems to be testing the whole range of the compass. Many stores now carry weather vanes, and some libraries have books illustrating how the vanes are made.

The "weathercock" helps us to be weatherwise, instead of otherwise.

Four Doweled Sticks Tell Forest Fire Danger

Out in the depths of the Snoqualmie National Forest, a ranger picks up four sticks doweled together. He weighs them. The weight is 105 grams. "Whew!" he says to himself, apprehensively.

The weight of the sticks means that the forest is like tinder, and could burst explosively into flame at the drop of a spark. The dry forest duff crackles underfoot as the ranger walks to his little "weather station" to measure the relative humidity with a psychrometer. The heat boils down from the burning sun. The humidity is 20 percent. That's bad. A reading of 30 would be dangerous, but 20—"Whew!"

A rain gauge stands beside the ranger. He pays no attention to it, except to avoid touching the blistering hot metal. He can't remember when it rained last.

The ranger replaces the sticks that gave him a measurement called "fuel moisture content." They are suspended six inches above the ground over typical forest "floor." Each is a half-inch thick and 20 inches long. The four, together, weighed exactly 100 grams after being cut, doweled, and kiln-dried.

Now they weigh 105 grams. That means they contain five grams of water which the moisture-hungry wood has "stolen" from the forest air. The sticks are good indicators of the forest's wetness or dryness. Their present light weight reveals that the whole forest is in peril.

The ranger phones two other men in his district to get data from other "weather stations." It's the same story. Hot and dry. There is no wind, thank God, but these men of the forest know that tricky winds can spring up quickly and funnel furiously through mountain valleys.

With data penciled on a pad, the ranger goes to his office and calls the fire weather forecaster for the National

Weather Service. The forecaster tells him it's a bad situation and warns that lightning can be expected during the night.

"Whew!" the ranger answers. "We've stopped logging and closed the district to campers. With luck we could avoid a conflagration, but lightning is like a lash from the devil himself. A lightning storm can give you anywhere from one to twenty fires in a couple of minutes."

The weather forecaster nods in agreement. He knows the Snoqualmie Forest almost as well as his ranger friend. He is familiar with all of the other federal forest lands in the state, too. He has to be. It's his job.

He issues forecasts each day during the fire season for the national forests within his jurisdiction and for timbered areas controlled by the National Park Service or the Indian Service. Sometimes one forest will be wet while all others are dry, or the situation may vary from one section to another in the same forest.

It was Harold Lindquist, former fire weather forecaster in the Seattle forecasting center, who told me how the doweled sticks are used to indicate hazardous conditions in the forests. He was on the National Weather Service staff then and one of the busiest men at the station when the fire danger was high.

Degree Days Data Aids Fuel Forecasts

Has your fuel dealer told you about degree days? Probably not, because he's busy and on the go, but he uses degree-day information to gauge your fuel needs. You can use it, too, if you want to keep a close check on fuel consumption.

The degree-day figures, which the National Weather Service supplies, help oil dealers, gas companies, wood and coal suppliers, and others in the fuel business. They check degree-day data to determine how much fuel you may have used over a given period.

Those who deliver oil or other types of fuel to your house get to know your needs pretty well by studying degree-day figures. They know, without asking you, when to come with a new supply.

The degree day is a unit, based upon outside temperature. It is determined by subtracting the average temperature for the day from a base figure, 65 degrees. If the average outside temperature for a particular day was 25 degrees, the number of degree days (for that day) would be 40.

The average temperature for a specific day is determined by adding the maximum and minimum temperatures, then dividing by two. Here's a sample:

Maximum temperature, 50, minimum, 40. The total is 90. Divide by two and you get 45. Subtract 45 from 65 and you have 20 degree days.

Sixty-five degrees is an ideal base figure, because an average temperature of 65 outdoors would mean that virtually all sources of artificial heat would be cut off both day and night. It would normally be warmer than 65 indoors if the outside temperature was 65.

If the degree-day figure is large for the day, the week, or the month, your supplier of fuel knows you have used a lot of his product. Degree-day normals for the different seasons are particularly useful in helping fuel dealers plan ahead.

The home-owner can use the degree-day data to keep a close check on fuel consumption. If, for example, the total number of degree days for this month is the same as the total for last month, fuel consumption for each month should be about the same.

However, there are factors that can alter the situation. If your Aunt Minnie, old and feeble, comes for a visit she may want more heat than you need. You might increase your fuel consumption temporarily for a newborn baby, too.

Many oil dealers divide the total number of degree days for a specific period by the number of gallons placed in your tank at the last filling. This gives a figure, called the K Factor, which is used as a guide in helping them determine when you need more fuel.

The general K Factor is adjusted to a specific K Factor for each customer.

One dealer told me it usually takes two deliveries to get

an individual customer's K Factor. In general, this dealer figures that a 300-gallon tank full of fuel will last through 1,050 degree days in the area he serves.

Heating engineers, seeking a more exact K Factor, would take into consideration the amount of solar radiation received by your community, plus wind conditions and outside temperature. They also would measure heat loss in your home through windows, walls, and doors.

But the degree-day method based on average outdoor temperatures is good enough to be widely used in determining your fuel requirements.

Our normal accumulation of degree days at the Seattle-Tacoma International Airport is 5,185 per year, and this figure would be representative of most residential communities in this area.

Translating the Words of the Weather Service

Why does the weatherman use the term "Partly Cloudy?" Why not be optimistic and say "Partly Sunny?" What does the forecaster mean when he says "20 percent chance of rain," or uses some other percentage figure—30, 40, 50, or whatever—to indicate the probability of precipitation?

Those questions deserve answers because some people think the forecaster is unduly pessimistic at times, especially during our gloomy winter months, and others misinterpret forecasting terminology.

First, it should be pointed out that all major forecasting centers, like Seattle's, have more than one forecaster. Each major forecasting center is staffed around the clock, seven days a week, with meteorologists in three principal classifications: district forecasters, aviation forecasters, and public service forecasters. In addition, there are technicians, map plotters, and operators of the equipment that receives satellite pictures.

There is, of course, a head man whose title is meteorologist-in-charge. He takes the compliments or criticisms in stride and is, in truth, *The Weatherman*. He can walk down the street on a clear, sunny day and smile happily as one acquaintance after another says hello and adds, "Thanks for this weather. It's great! Keep it up!"

But on a gloomy day the greetings are not as cheery, and he hears comments like: "Lousy weather you're giving us. Can't you shut the faucet?"

Seattle's weatherman is Dr. Arthur Hull.

There was a time in the 19th century and early in this century when the Clerk of the Weather bore the brunt of ridicule or was showered with praise. The "clerk" was an imaginary functionary who supposedly controlled the sun, rain, clouds, and other elements. But the "clerk" lost out when the Weatherman emerged as an important figure.

Now for terminology:

Cloudy means that the sky is covered from seven tenths to ten tenths by clouds dense enough to obscure sun, moon, and stars.

Partly Cloudy—three tenths to seven tenths of sky covered by clouds. The term is interpreted by some to mean the sky is partly covered by clouds during all of the forecast period, and by others to mean the coverage is for part of the forecast period.

Partly Sunny is acceptable terminology for daytime forecasts, but obviously not for a period of twilight or darkness. So if the acceptable degree of cloudiness prevails (about one quarter to three quarters of sky coverage), and if the forecast period coincides with the time the sun is shining, the words, **Partly Sunny,** probably would be used. However, if the forecast period starts in darkness and ends soon after sunup, the forecaster probably would say **Partly Cloudy.** This is done to prevent introduction of another term which might be interpreted as meaning something different, even though the expressions are interchangeable. The same reasoning would prevail if the forecast period began in daylight and ended at night.

Mostly Cloudy is a term used when cloudiness will be subject to some variability in amount or location. It indicates a condition which is expected to predominate.

Variable Cloudiness is intended to describe an irregular sky condition, such as one in which bands of clouds drift across the sky. It is used when the forecaster expects cloudiness to increase and decrease several times during the forecast period.

Some Sunshine implies an indeterminate amount which may vary in location or time.

Clear is reserved for a sky that will be free, or virtually free, of clouds. The sky should not be obscured by more than two tenths to quality.

Fair means no likelihood of precipitation, with clouds covering less than four tenths of the sky. Low clouds are meant in such a forecast because high, thin clouds through which the sun, moon, or stars are visible do not convey the impression of cloudiness to the public.

One other particularly difficult weather term deserves some attention—the **Occluded Front:**

The TV weatherman steps forward briskly, pointer in hand, and aims it at a map of the United States. He shows you the "highs" and "lows," reveals where tongues of cold air are inching down from the arctic and tells where rain, snow, and hail are falling.

Then, having emphasized how lucky you are to be living here instead of "there"—where tornadoes whirl, the snow piles deep and ice storms create havoc—he draws a picture of a low-pressure system off our coast. Better yet, he may show you a picture of clouds from a weather satellite.

"There's an *occluded front* out there," he says, tapping the spot with his pointer. "It's going to bring us rain and some wind, but there'll be breaks in the overcast. So, we can expect a little sun, too."

You hear the references to wind and rain and remember the warning about raincoat weather. You've gotten the essentials of the forecast, of course, but what about that word, *occluded?*

If you don't know the meteorological meaning of occlusion, but have wondered about it, you're not alone among weather fans.

First, let's get acquainted with the word. To occlude means to shut, close, or obstruct. You can occlude light or a knot in a tree trunk. You can create an occlusion in your mouth by bringing your upper and lower teeth together.

I asked a National Weather service forecaster about our wintertime weather, and he gave a lucid and interesting answer which involves occlusions. He said:

"Typical highs and lows and simple warm and cold fronts account for only a small percentage of the weather changes experienced in this area. Most of the time, in this region, pressure systems and temperature fronts are interfered with by the topography and modified so much they are almost unrecognizable if the wind and weather we experience is compared with what is supposed to occur, according to the classical models.

"It is true that changes in barometric pressure around highs and lows, and changes in air masses with the passage of fronts, do occasionally produce weather changes. But if you expect them to follow the textbook pattern you may be misled."

Our principal weather breeder from October through March is the Pacific Occluded Front. We may presume that an occlusion is approaching if the sky gradually darkens, the barometric pressure falls, and the wind increases slowly, provided there is no drastic change in the temperature.

When those developments occur, it may be only a matter of hours before rain begins.

"Seldom will the rain be heavy, especially in the lowlands and valleys," he said, "but it is virtually certain to last for several hours.

"Now watch the barometer. If it stops falling, you may expect a small change in wind direction. The wind will take on more of a gusty character.

"The rain will stop temporarily, after falling for a number of hours. The sky will become brighter and the visibility sharper. But any sunshine that may appear will not last long. Rain will come again, this time in the form of showers.

"The temperature will have dropped a few degrees but hardly enough to call for a change of clothing. As the showers gradually lessen, you can assume that the occlusion has passed on to the east."

We have an occlusion coming or going almost continually in the late fall, winter, and early spring. An occlusion forms in the center of a low pressure area. The winds circling a low (moving counterclockwise) will bring together a cold front and a warm front around which the low has been generated.

Wherever the two fronts are joined, an occlusion has occurred.

Now for **Probabilities of Precipitation.** A forecast calling for a 60 percent chance of rain does NOT mean that it will rain only 60 percent of the time. Nor does it mean that only 60 percent of the forecast area will get rain. "Chance of rain 60 percent" means that rain can be expected six times out of ten when the same atmospheric conditions prevail. Such a forecast is based upon the weatherman's experience and knowledge, both of which are important.

The forecaster's knowledge of meteorology and associated sciences would be less than adequate if he wasn't thoroughly acquainted with his forecast region—with all of its peculiarities and oddities of weather. It is a "must" for the forecaster to know the terrain of his area. He must understand how mountains slow down, divert, or block storms, how valleys channel the wind, and how water will affect temperatures.

The forecasters in Seattle must know through which gaps cold waves move into the region, how the Cascade Mountains protect us from most of summer's intense continental heat and much of winter's chilly air. They must be familiar with the storm paths and the customary speed of weather disturbances moving inland from the ocean.

They have to know where, how, and when our fogs develop. They must analyze data from upper air soundings (temperature, barometric pressure, and wind velocity) in order to determine whether upper level "lows" are sneaking into the area to disrupt surface conditions. And they have to spend considerable time studying maps of surface data. The photographs of clouds, especially "portraits" of whirling storms taken by cameras on weather satellites are a big help.

The time may come when the world will have enough

weather reports from everywhere, including eyes in the sky, to take all the sneakiness out of the weather. But that time isn't yet at hand. Meteorology hasn't become an exact science, but our National Weather Service is doing a pretty good job of solving the major twists and turns of earth's blanket of air.

As one of our television meteorologists occasionally points out to his listeners:

"The forecasts issued by forecasting centers of Oregon, Washington, and British Columbia are almost always correct when they advise us that a storm is approaching or that fair weather is on the way. It's the timing that's wrong, at times.

"And the timing is occasionally wrong because the atmosphere is an enigma wrapped up in a mystery. A storm moving landward from the Pacific Ocean may suddenly slow down and creep inland. Or, conversely, a slow-moving weather front might pick up speed and arrive sooner than anyone expected."

Lack of data from the vast Pacific Ocean is the major handicap faced by the Northwest's forecasters. They get some data from the Canadian weather ship, which is stationary several hundred miles off the British Columbia coast, and from ships plying the Pacific Ocean, but there aren't enough weather observers on the Pacific to give them all the help they need to be more specific.

Many a storm has bypassed the weather ship and other vessels and smashed full tilt through the storm gates of Washington, Oregon, and British Columbia.

The Long, Really Long-Range Forecast

Has our Pacific Northwest climate changed in the past few centuries? Yes indeed! It has varied from cool to warm to cool again, with some moderate in-between weather, according to a study made by the late Prof. A.E. Harrison in the 1950s.

The professor, who taught electrical engineering at the University of Washington, was a weather hobbiest and mountain climber, and the glaciers of Washington fascinated him. So he took up glaciological research and made some interesting discoveries. His studies indicated that cold weather was experienced here as follows: just prior to 1750, just before 1850, and then from 1870-1880, 1892-1905, 1932-1936, and 1945-1959.

Prof. Harrison had the benefit of climatological studies, of course, but his older listings emerged largely as educated guesses after years of probing into, under, and around ice masses, and then "reading" the ancient history disclosed by glacial deposits of broken trees, stones, earth, sand, and debris. The notable warm periods showing up in those studies extended from 1884 to 1891, and from 1920 to 1931, as well as in 1952 and 1958.

The professor said he was particularly grateful to Arthur J. Johnson of the United States Geological Survey, whose disclosures of glacial growth in the Northwest from 1946 onward captured the attention of glaciologists throughout the world.

A number of other scientists have helped fill in missing links in the climatological picture of olden times. Most of the studies have been confined to Nisqually Glacier and other ice masses on Mount Rainier and Coleman Glacier on Mt. Baker.

Nisqually Glacier, which flows down the south slope of Mt. Rainier, reached its greatest advance in the last 1,000 years somewhere between 1832 and 1843. The terminus then was a little over a mile down-valley from Nisqually's terminus 135 years later. That farthest advance in 1,000 years would have put the terminus about 800 feet below the new bridge over the Nisqually River between Longmire and Paradise Valley.

That determination was made by studying trees in seven areas on and near the terminal and lateral moraines left by the glacier after its big push in the first half of the 19th century. (A moraine is a deposit of rocks, earth, and other glacial debris.) The oldest trees in the areas studied are now from 130 to 136 years of age. They started to grow when the moraine became stable, following glacial retreat. If five years are added to those ages, to allow for stabilization of the moraine, the debris has remained fixed for about 138 years. So, for convenience of numbers, scientists are calling it the 1840 moraine. The oldest trees among the debris would have germinated in about 1842, according to the age count.

Down-valley from the 1840 moraine there are trees as old as 316 years, and the humus layer on the forest floor, which includes rotten wood, represents 200 additional years of continued growth.

In addition, there is charcoal beneath the humus which represents a still earlier generation of trees from 100 to 200 years old. Altogether, between 600 and 700 years of forest growth, except for interruptions by fire, are accounted for down-valley from the 1840 moraine.

Several layers of organic matter and pumice even farther down Nisqually Valley, but still within Rainier National Park, are believed to represent 300 or more additional years of ice-free terrain. Thus, the time that elapsed between the last major glaciation of Nisqually Valley and the advance marked by the 1840 moraine probably was at least 1,000 years. And the period might be considerably more than one millenium.

The 1840 moraine of Nisqually Glacier is from 10 to 20 feet high and 25 to 50 feet wide. It is composed of angular blocks and is treeless at many places, thus forming an open lane between the old forest and the younger trees.

Studies still going on in other glacial valleys within Rainier National Park show prominent moraines corresponding in time to the 1840 moraine of Nisqually Glacier.

The moraine and botanical evidence below the Emmons Ice Mass on Mount Rainier's northeast slope indicate two other significant advances of the ice, one in 1745, the other in 1895. The 1745 moraine marks the maximum advance for Emmons Glacier in modern times. Tahoma Glacier on the mountain's southwest flank made a tremendous forward surge early in the 17th century, and again in the early 19th century. The dates of those advances have been fixed at 1635 and 1835.

All of Rainier's glaciers have receded, then advanced,

then receded in recent decades. All in all, the mountain's ice masses are smaller now than at the turn of the century.

But no one knows for certain whether the earth is warming up permanently. Back in the mid 1950s the world's mean temperature was 1.5 degrees higher than it was at the turn of the century. But since then Europe, Canada, the United States, and Russia have experienced some of the worst winters in more than a hundred years.

It's difficult to get a definite fix on climatic trends because we're always in some kind of a weather cycle, and cycles occur within cycles. Many a long warm span has had a few cold years and a lot of cold spells were broken temporarily by spurts of warmth.

Many scientists claim that carbon dioxide, the gas you breathe out each time you exhale, has produced a "greenhouse" effect around our planet. Man's smokestacks, engine exhausts and trash fires are the chief sources of carbon dioxide pouring into the atmosphere.

Carbon dioxide helps retard the radiation of heat into space from the earth. Water does, too.

It has been estimated that the amount of carbon dioxide hurled into the atmosphere by mankind's fires alone has increased from a half billion tons in 1860 to nearly 11 billion tons per year now. It could reach 20 billion tons annually by the end of the century.

Will man be the cause of the next major climatic change? Only time can tell us.

What to Expect, Usually

January is Unusual, Usually

IN my youth I thought our January weather here in Western Washington an unkind offering from the eternal verities. The opinion developed before I had seen any of January's weather elsewhere in the nation.

I changed my mind after trudging through deep snows on the East Coast, getting chilled to the marrow in Midwest blizzards, and experiencing the discomfort of blazing heat soon after New Year's Day in California's Imperial Valley.

It was always nice to come home, and only once—following my retirement—was I tempted to give the warm and sunny Southwest a chance to prove me wrong. The experiment lasted less than a year. I missed our Northwest's rain, the tall trees, the sparkling lakes, our rushing streams, and that jewel among the world's waters, beautiful Puget Sound. So I came back to what I consider "God's Number One Country."

My army service during World War II took me to various parts of the United States and Canada. Wherever I went, I told my soldier friends about "God's country out West," and a buddy from Oregon added his superlatives to mine. After the war, one New Englander came to visit me and my Oregon friend.

McCAUSLAND

"I wanted to see for myself the wonderland you fellows described in such glowing terms," he said. After numerous tours in both Oregon and Washington, he said, "I agree with you fellows. This is surely part of God's country; the other part is in Vermont and New Hampshire. They can give all the area in between back to the Indians."

I don't agree with his assessment of America's heartland. There are innumerable good climates and scenic areas between the Rocky Mountains and my New England friend's Green Mountains. But for all-around climatic comfort, I'll take the Pacific Northwest and so will my Indian friends who were born and raised here. They, too, prefer "God's Number One Country."

I'll take our Januarys in preference to the Januarys in any other part of the nation, and so will thousands who have moved to this region to escape frigid winters beyond the Rockies. Sure, we get January frosts and some ice and snow, but the chill never lasts long. The maritime air off the ocean is warm most of the time, even in the dead of winter, and the grass stays green. Seattle's average temperature in January is 38.2 degrees Fahrenheit. The average maximum temperature is a pleasant 43.4 degrees, and the mean minimum mercury reading is only 33.

It's a good idea, though, to keep the earmuffs handy. We've had a foot or more of snow in some Januarys, and arctic cold waves have, on a few occasions, had enough push to surge over the Cascades into the Westside lowlands or to pour southward out of British Columbia's western valleys.

Most of the cold weather in Western Washington comes after December 21, the first day of winter, and before February 1. The average temperature begins to climb late in January and keeps on rising until August. January frosts are common, but major snowstorms are rare in January or any other month.

I like to see the snow gleaming on our Cascade and Olympic Mountains, and I like knowing that that's where it falls most of the time.

Perhaps because the unusual is the usual for January, there are dozens of January proverbs that attempt to prepare folks for the worst.

"A cold January, a feverish February, a dusty March, a weeping April and a windy May presage a good year and gay," French proverb.

"Always expect a thaw in January," says grandpa.

"January will freeze the pot by the fire," Old English. "As the day lengthens, the cold strengthens (in January)."

Those are a few of the old sayings about January's weather.

According to ancient Saxon lore, the trend of the year's weather is influenced by the day of the week on which January 1 falls.

Monday, a severe confused winter, followed by a good spring but a windy summer; Tuesday, a dreary and severe winter, a windy spring, a rainy summer; Wednesday, a hard winter, a bad spring, a good summer; Thursday, a good winter, a windy spring, a good summer; Friday, a variable winter, a good spring and summer; Saturday, a snowy winter, blowy spring, rainy summer; Sunday, a good winter, windy spring, dry summer.

Another proverb tells us: "March in Janiveer, then Janiveer in March, I fear." That one seems to hold true for these parts. We usually get a warm spell in January and some coolish, January-like weather in March.

The Scots have a saying: "A January spring is worth naething." How true!

If January calends be summerly gay,
It will be winterly weather till the calends of May.

The name, calends, is a reference to the first of the month in the old Roman calendar.

A Swedish proverb says, "It will be the same weather for nine weeks as it is on the ninth day after Christmas."

Another old belief, imported from Europe, is that "the first three days of January rule the oncoming three months." How about that?

It also has been said that "the last 12 days of January rule the weather for the whole year."

The Greeks say, "If January could, he would be a summer month." Well, he is in the southern hemisphere.

Warnings about warm spells in January are numerous:

"January warm, the Lord have mercy!" (Because later frosts may kill premature new growth.)

"When gnats swarm in January, the peasant becomes a beggar."

There's a verse about St. Vincent's Day, January 22:

Remember on St. Vincent's Day,
If that the sun his beams display,
Be sure to mark his transient beam,
Which through the casement sheds a gleam;
For 'tis a token, bright and clear
Of prosperous weather all the year.

St. Paul's conversion day (January 25) has been associated with the weather too. On the Isle of Man, 'tis said:

Paul's Day stormy and windy,
Famine on earth, and much death on people;
Paul's Day beautiful and fair,
Abundance on the earth of corn and meal.

A French saying goes like this: "If St. Paul's Day be fine, the year will be the same."

Theophrastus, a Greek philosopher and naturalist who lived before Christ, said: "If the winter sets in early it will close early, but the spring will be fair; if the contrary, spring also will be late. If the winter is wet, the spring is dry; if the winter is dry, the spring is fair."

A Northern European rhyme says: "The blackest month in all the year is the month of Janiveer."

February: Lull Between Storms

When the cat in February lies in the sun, she
she will creep behind the stove in March.
When the north wind doth not blow in February,
it will surely come in March.

There's a lot of truth in that proverb, a European saying that applies here, too.

February's calm is simply a lull between storms. There's

still feudin' and fussin' going on in the atmosphere, particularly over the Aleutian Islands, where frigid arctic air meets warm (or at least warmer) ocean air. The Aleutian low-pressure system, a breeder of storms, is still a potent force because of those temperature contrasts. And we are the recipients of many February storms.

Isolated fine days in Surrey, England, are known as "weather breeders," and the French say: "February singing never stints stinging." A proverb from Cornwall tells us that: "A February spring is not worth a thing."

There are good reasons for those proverbs, the principal one being that premature growth of vegetation may occur, only to be followed by killing frosts in early spring.

Among other Old World proverbs are these: "Februeer doth cut and sheer," England.

"February rain is as good as manure," France.

In February, if thou hearest thunder,
Thou will see a summer's wonder.

Sioux Indians.

"February, shortest and worst of all the months," France.

If in February there be no rain,
'Tis neither good for hay nor grain. *Portugal*

If February gives much snow, A fine summer it doth foreshow. Early American.

February's weather here usually includes one or two mild spurts of warmth to remind us that spring is coming. It's not unusual for us to get balmy weather early in the month, with a chilly spell following.

February, 1968, was Seattle's warmest February since the National Weather Service began taking observations at Seattle-Tacoma Airport in 1945. The average temperature during that month was 48.5 degrees, which was 6.2 degrees above normal.

March—The Unpredictable

There's a saying on the island of Kithnos in Greece that

"March was so angry with an old woman for thinking he was a summer month that he borrowed a day from his brother, February, and froze her to death."

The proverb has its merits in pointing up the fact that March weather is unpredictable. The month may bring rain, snow, hail, sleet, frost, or rime. Yet it can, and sometimes does, produce some of spring's most pleasant days in our Pacific Northwest.

In Northumberland, England, a county bordering on Scotland, 'tis said:

March yeans the lammie and buds the thorn,
And blows through the flint of an ox's horn.

The French say "only fools go barefoot in March," and, "March never has two days alike."

The weather lore of Scotland tells us:

A peck of March dust and a shower in May
Make the corn green and the fields gay.

The Scots also say:

March wet and windy
Makes the barn full and finnie.

(Finnie refers to the feel of the grain, indicating quality.)

The Germans contend that "snow in March is bad for fruit and wine," and that "Thunder in March betokens a fruitful year."

Fortunately, we don't get big hailstones here. Our thunderstorms are strictly second-class compared to those in the Midwest, where hail may wipe out crops in a matter of minutes.

We don't get the Midwest's smashing rains, either (as a rule), although the late Portus Baxter, Seattle pioneer, used to tell us that: "The rains here seem to be heavier now, and the winds more furious, than in the days of my youth."

We are inclined to believe that neither the rains nor the wind are any different now. The pioneers simply had more protection from wind-driven rains and violent gales, thanks to the forests that rimmed our little village in the long ago.

We still get a lot of the drizzly "Oregon mist" that old-timers remember.

Seattle's mean precipitation (both drizzle and downpour) for March is 3.61 inches. The city's wettest March in history was March, 1879, when the skies wept almost continuously, yielding a sea of water totaling 11.92 inches. The information for that particular month is based upon records kept for the government prior to the establishment of a National Weather Service office here, but there is no reason to doubt the observer's accuracy. The wettest March that I can recall was March, 1950, when rainfall totaled 7.23 inches in downtown Seattle.

Seattle's average March temperature is 44.1 degrees Fahrenheit, which is sufficient to cause fruit and flower buds to swell in joyous anticipation of blossoms in April or May.

The average March snowfall at Paradise Ranger Station on Mt. Rainier (elevation 5,550 feet) is 32.2 inches, but in March, 1971, the snows piled to a depth of 215 inches there, and a year later the March snows exceeded 109 inches in depth. (Mt. Rainier's biggest snow years in history were those of 1970-1971 and 1971-1972. The snow year is the period from July 1 to June 30.)

Seattle, whose weather is typical of many lowland cities west of the Cascade Mountains, gets little, if any, snow in March. Portland, Oregon, is also mostly snow-free in March. But March has, on a few occasions, been cold and snowy in both cities. March, 1951, was one of those exceptional months. The snow totals in that wintry March were 12.9 inches at Portland and 18.2 inches in Seattle.

In general, though, March will offer some gentle spring weather in between winter's last gasps, especially along the coastal strips of Oregon, Washington, and British Columbia, and in the Puget Sound lowlands and Oregon's Willamette Valley.

At the time of the vernal equinox, on or near March 21, days and nights are virtually equal all over the world, and that in itself is a plus for the Northern Hemisphere. From then on, until June 21, our days are longer than the nights, and spring growth will receive encouraging light and warmth.

Some April Showers

April's anger is swift to fall,
April's wonder is worth it all.

Sir Henry Newbolt

The deep blue eyes of springtime usually smile benignly upon us in April from the skies above, meadows below, and all the woodlands.

We see the hostess of the sky, the moon, more often at this season as the weather clears. Spring flows through the valleys and over the hills to open the violets, rhododendrons, camellias, daffodils, and other flora.

And if your pulse doesn't quicken at the sights and sounds of the season—the bursting buds, the croaking frogs, and the predawn concerts of the birds—then you'd better snap out of your own winter "sleep" and poke around outside.

Take another look at those lines by Sir Henry Newbolt. April is like that, angry one day and sweet as syrup the next. Sir William Watson, a contemporary of Sir Henry, must have had April in mind when he wrote:

Trees in their blooming,
Tides in their flowing,
Stars in their circling
Tremble with song.

And we'd like to think that William Wordsworth wrote the following lines in spring:

One impulse from a virgin wood
May teach you more of man,
Of moral evil and of good
Than all the sages can.

One of the proverbs about spring says: "A cold April the barn will fill." Could be! But much depends on the weather during the following months. If April is continually cool, most new growth will come slowly. But a warm May can bring about bountiful yields in the gardens.

A killing frost is more to be feared in April than anything else in the way of weather, as nature must then make a fresh start. But killing frosts usually do not come in April in the western lowlands of Washington, Oregon, and British Columbia.

Spring puts on her most lavish floral displays west of the Cascade Mountains during the last two weeks of April and the first half of May.

Seattle's average April temperature is 48.7 degrees Fahrenheit, a figure that is fairly representative for most Puget Sound communities. The mean monthly precipitation at Seattle-Tacoma International Airport totals 2.46 inches. But we've had many an April that was drier, and some considerably wetter. That's April for you, a transitional and unpredictable month.

Early in the month it's always touch-and-go whether the storm track will move north like the birds or remain a while in our backyard, but stormy or not we can look for an increase in average temperatures.

Portland, Oregon's mean monthly temperature jumps from 45.7 degrees in March to 50.6 in April, and the average precipitation drops from 3.64 inches in March to 2.22 inches in April.

Both Eastern Washington and Eastern Oregon begin to warm up in April, too, and so do the interior valleys of British Columbia.

Yakima, Washington, usually gets more daytime warmth than Seattle in April, but the nights are cooler than those west of the Cascade Mountains. However, the warmer daylight hours result in an average April temperature of 49.5 degrees, just a bit higher than Seattle's average.

But it doesn't take long for all areas east of the Cascades to surpass Westside communities in springtime warmth and dryness.

April is still a snowy month in the mountains though. Paradise Ranger Station on Mt. Rainier (elevation 5,550 feet) can expect more than 50 inches of snow during the month and has had as many as 143.5 inches of snow in April. That mammoth pileup of the white stuff fell in April, 1972.

So it goes in April. It's a spring month, yet winter retreats reluctantly.

Mild May

What are the May weather prospects in the Pacific Northwest? A proverb still extant says, "Don't praise the month till it's ended." We won't in any year, yet May merits some kind words on the basis of probabilities alone.

I can vouch for the splendor of our outdoor gardens, both wild and cultivated, in early May. It's a wonderful time of year. The season's warmth takes a big spurt forward in May, thanks to longer days and the gradual shrinking of that mammoth cloud, wind, and rain factory, the Aleutian low-pressure system, in the North Pacific Ocean. The storms don't cease entirely, but they are much less severe than those of winter and early spring, and the storm track tends to shift northward, giving Washington better weather.

The increasing length of the day (the time between sunrise and sunset) is a major factor in warming us in May. Seattle's day, for example, is 14 hours, 28 minutes long on May 1, and 15 hours, 40 minutes long on May 31. To those figures you can add long stretches of morning and evening twilight, which help shorten the nights.

Seattle's mean temperature in May is 54.9 degrees Fahrenheit (data for Seattle-Tacoma Airport) and precipitation averages 1.70 inches. Temperatures in and near the city have climbed into the 90s on a few occasions; it doesn't happen often. The average of the maximum temperature for May at Sea-Tac is a pleasant 64.1 degrees Fahrenheit; the average minimum is 45.6 degrees.

The all-time May high at the airport was 93 on May 21, 1963. The high the day before was 92. That particular May was one of the warmest and driest in the city's history. There were 26 days without measurable precipitation at the downtown office of the National Weather Service, equaling the record established 35 years previously. The mean temperature that month was 58.6 degrees downtown and 57.7 at the the airport.

May can turn wet and cool, too—as in 1960 and 1961—when the Pacific high-pressure cell was too far offshore to prevent weather disturbances from slipping down the coast from the Gulf of Alaska. But that's the exception, not the rule. Most Mays are sunny, pleasant, and fairly dry. In both 1960 and 1961 the rainfall total exceeded three inches.

May freezes are rare in most areas of Western Oregon and Western Washington. The only one of any consequence in modern times occurred on May 1, 1954, when the mercury dropped to 28 degrees at Seattle-Tacoma Airport. Many a tender flower was nipped by that frost.

The chance of such a late cold snap here is almost nil. In fact, there's only a slight chance of 28-degree weather occurring in Seattle's outlying areas up to April 4. Thereafter, the betting would be overwhelmingly against freezing temperatures.

Climatological records show that May often is sunnier than June in the greater Seattle area. May's rainfall is only slightly over the mean for June, and many times June has been the wetter month.

Seattle's driest Mays—each with less than one inch of rain—were those of 1884, 1888, 1898, 1904, 1907, 1914, 1917, 1920, 1924, 1928, 1932, 1935, 1946, 1947, 1950, 1952, 1956, 1958, 1963, 1964, and 1972.

May, 1928, with 0.31 of an inch of rain, was Seattle's driest, yet the last two days of that month were so chilly that tender vegetation suffered damage in many areas. Cranberries in Southwest Washington were hard hit by frost on May 31, 1928.

Seattle's second driest May came in 1904. Precipitation totaled only 0.31 of an inch. Frost also was recorded in that month during a dry spell. Damage occurred in strawberry fields and some truck gardens in the surrounding area.

For near-perfection in May weather in Western Washington, I give the nod to May, 1958. The downtown weather station in Seattle listed 22 days without measurable rain, and maximum temperatures were in the 80s on seven days. Twelve additional days had mercury readings in the pleasant 70s, and the average temperature was a balmy 62.3 degrees, making it the city's warmest May in history. The maximum reading was 87 in downtown Seattle and 89 at Seattle-Tacoma Airport on the 18th.

On the east side of the mountains, the mean May rainfall totals just a little over a half-inch in the Yakima Valley and the Wenatchee area and close to one and one-half inches in the Spokane and Walla Walla regions.

Clouds! June in the Northwest

June 21 is the date of the summer solstice, the longest day of the year. The sun rises in the Seattle area on that date at 4:12 a.m. and sets at 8:10 p.m., Pacific Standard Time (add an hour for daylight time). Each day thereafter is shorter by seconds.

Astronomically speaking, the 21st of June is the beginning of summer, yet it's the date on which the year begins to die. All of Nature's efforts through the remainder of summer and into the fall are bent on ripening the fruits of the harvest, and strengthening the bodies and wings of springtime's baby birds in preparation for their migrations.

The sun shines directly down upon the Tropic of Cancer at 23 degrees 27 minutes North Latitude on the date of the summer solstice. Thereafter, the ancients believed, the sun travels south until it reaches the Tropic of Capricorn at 23 degrees 27 minutes South Latitude on December 21. Then the year commences anew.

Actually, of course, the sun doesn't travel north or south. That so-called journey of the sun comes about as the result of the earth's tilt. Our Northern Hemisphere is tilted toward the sun in summer and away from it in winter.

June is generally warmer, but a little cloudier than May in the Puget Sound region. That's because of fast warming on the east side of the Cascade Range. The extra heat in Eastern Washington causes thermal "lows" that suck in air from the West. The air drawn away from Seattle is replaced by moist, cloud-laden air off the ocean.

By mid-July, if not sooner, the land in Western Washington is hot enough (on its own) to exert a drying influence upon ocean air. Result: fewer clouds, more sun, and better picnic weather.

Fogginess in summer on the Washington coast is the result of another phenomenon. The fog develops when Pacific air (relatively warm) passes over a cold current of water. This summertime current is found, in most years, near the outer coast of Vancouver Island. It exists because persistent northwest winds churn the ocean, causing an upwelling of chilly water from the depths. Fog is a natural result when warm air meets cold water, or vice versa.

July Brings Shirtsleeve Weather

Then came hot July, boyling like to fire.
Edmund Spenser

Our Julys in Western Washington and the west coast of British Columbia don't ordinarily "boyle like fire," nor do the Julys in England where Spenser lived. But we get an occasional burst of July heat on our side of the state. Fortunately, though, our warmest summer days are marked by low relative humidity, a blessing not shared by most regions east of the Rocky Mountains.

Seattle's highest temperature was 100 degrees Fahrenheit, recorded first on July 16, 1941, and equaled on June 9, 1955. Both of those readings were made when the National Weather Service recorded official temperatures at the old Federal Office Building on First Avenue at Madison Street.

July usually is the warmest month throughout the Seattle area, but the month hasn't always brought the highest temperatures. For example, the all-time maximum temperature at Seattle-Tacoma International Airport was 99 degrees on August 9, 1960; this compares with Sea-Tac's June high of 96, recorded on that day in 1955 when the mercury climbed to 100 in downtown Seattle. The highest thermometer reading at Sea-Tac in July was 97 degrees on July 12, 1961.

If you are a newcomer in this area, I hasten to inform you that summer temperatures in the 90s are the exception, not the rule. The average maximum temperature for July at Sea-Tac is a pleasant 75.1 degrees.

Summer brings shirtsleeve weather, that's for sure, but not the muggy, debilitating weather that hits large sections of the Deep South, Midwest, and the East Coast during the warm season. And it's a rare summer night in the Northwest when a light blanket isn't needed for sleeping comfort.

The driest time of the year in virtually all communities on the west side of the Cascade Mountains usually coincides with the last ten days of July and the first week or two of August. Most of the major outdoor festivals are held at that time of year under blue skies and a smiling sun.

Eastern Washington, Eastern Oregon, and numerous places in the valleys and on the plateaus of British Columbia, east of the coastal lowlands, get sizzling hot in the summer. But even in those super-warm regions the relative humidity is low on hot days.

The big farms, orchards, and cattle ranches of Washington, Oregon, and British Columbia are on the east sides of the central mountain chains, where summer's heat is an asset and irrigation takes the place of nature's moist offerings from above.

Another English poet, Lord Byron, born in 1788, more than 200 years later than Spenser, spoke disparagingly about the climate of his native land. He said the English winter "ends in July, only to recommence in August." We've heard similar comments about the climate here, made by those who look upon a few days of rain as a disaster. We DO get generous rains, but it's mighty hard to beat a balmy July day in the Northwest. So when it sprinkles in summer, take heart; we haven't yet had a year without a summer.

August—Sun for September's Harvest

"August ripens, September gathers in; August bears the burden; September the fruit," Portuguese proverb.

Observe on what day in August the first heavy fog occurs, and expect a hard frost on the same day in October.

When the dew is heavy in August, the weather remains fair. Thunderstorms in the beginning of August will generally be followed by others all month.

These are samples of August proverbs handed down through the centuries. The first is simple summer logic; the second, hogwash; the third, partly true (the first half).

If you have a garden and fruit trees, August's sunshine will ripen your corn, redden your tomatoes, and fatten the fruit, provided water is available too. So August truly bears the burden for September's harvest.

But an August fog has nothing to do with October frost. Nor does one thunderstorm trigger others over a span of weeks.

Some places in the United States might have thunder and lightning every day in August, whereas other sections could have none. The Pacific Northwest has few thunderstorms at this season because of atmospheric stability.

Thunderstorms are most common (in central Florida, for example) where intense heat and unstable air create wild summer weather.

The proverb concerning dew is just about 100 percent right. If the grass is wet with dew on a summer morning, expect fair weather.

Let's have a look now at some of the highlights of August weather in past years:

August 1, 1954—The snows of winter and spring were late in melting at Paradise Valley, Mt. Rainier, elevation 5,550 feet. Sixteen inches of snow remained on this date.

August 4, 1961—A sizzling hot day in Eastern Washington. Both Spokane and Walla Walla experienced their warmest days; 109 at Spokane, 113 at Walla Walla.

August 5, 1961—A temperature of 118 degrees was recorded at Ice Harbor Dam in Walla Walla County, equaling the all-time high for Washington. The first 118-degree reading was made on July 24, 1928, at Wahluke in Grant County.

August 8 and 9, 1960—Seattle's warmest August days; 97 degrees on both dates at the downtown office of the Weather Service. Boeing Field measured 99 on the 8th and 100 on the 9th.

August 10, 1898—Oregon's highest temperature, 119

degrees, was recorded for the second time in the same year. The record was established on July 29 at Prineville, and equaled on August 10 at Pendleton.

August 14, 1933—The great Tillamook Burn in Oregon, this century's most spectacular and devastating forest fire, had its beginning in Gales Creek Canyon when a donkey engine huffed, cables pulled taut, and a giant Douglas fir dragged over a tinder-dry windfall, throwing sparks to the forest floor.

The temperature that day at Forest Grove, Oregon, a city on the eastern fringe of the fire, was 97 degrees. The next day Forest Grove's high was 103.

Ten days later the fire literally exploded along an 18-mile front, hurling flames thousands of feet into the air and smoke 40,000 feet aloft.

The great burn covered 311,000 acres, and of that total 240,000 acres were engulfed in 24 hours, August 24 to 25. Another 100,000 acres of timber, brush, and slashings went up in flames at the same time in the same general area.

August 14 and 15, 1950—Seattle's heaviest August rainfall in 24 hours, 0.79 of an inch.

September—The Golden Month

September is a golden month in the Pacific Northwest. Ponds lie calm and clear, the air is still, the haze gives a golden glint to the harvest moon, and new snow begins to whiten the highest mountains. The sun, fading southward for a summertime rendezvous with lands "down under," reaches the equator on September 21, the date of the autumnal equinox. At that time, as on the date of the vernal equinox in March, day and night are virtually equal throughout the world.

Our delectable blackberries, strawberries and raspberries, a summer treat, are long gone by September, but the blueberries and huckleberries are plump and tasty. Vegetables still come fresh from the garden, and fall flowers are at their peak.

Going back through National Weather Service records since the establishment of an office here, I found that June in the Seattle area was warmer than September in the majority of the years during the 1890s and again in the 1920s, 1930s, and 1940s. The average temperatures were a standoff for those months in the 1950s, and nearly a standoff in the 1960s, with June slightly ahead. September took the honors through most of the 1970s.

Actually, there's little difference in the warmth of June and September in the Northwest. The mean rainfall for each of the two months is less than two inches.

The three fall months—September, October, and November—usually are warmer than the spring months—March, April, and May—and there is a good meteorological reason for that.

The oceans are responsible in large measure for our weather, and in the fall the seas are considerably warmer than they are immediately following winter. So we benefit from the Pacific Ocean's stored-up heat.

Land surfaces in the northern hemisphere ordinarily reach their peak of warmth in July (or August at the latest), but the oceans don't reach their highest temperatures until late August or early September, as a rule. That's because the land heats up and cools off faster than water.

Ocean temperatures in Willapa Bay on the Washington coast are illustrative of the ocean's slow warming. The low for the year, averaging 44 degrees, is usually reached in February, and the high, averaging 63.6, is registered in August, but it's the late August readings that count the most in achieving that peak. Thanks to slow cooling, the average temperature in the bay is still in the 60s in September.

Seattle's Elliott Bay also is warmest in late August. Average temperatures for August and September are 56.1 and 55.4 degrees respectively.

But the peak of summer's warmth here in the air over land is usually reached between July 24 and July 31 when the average daily temperature stands at the year's high, 67 degrees.

Though it cools more rapidly than the oceans, the land is a storehouse of heat, too. That's why summer's warmest months are July and August, even though the sun reaches the end of its northward journey on June 21. Summer's heat keeps pouring into the receptive earth, and for a long time after the summer solstice we are the beneficiaries of accumulated warmth.

The same process of temperature lag works in the winter to make January colder than December even though our hemisphere's tilt away from the sun is greatest on December 21.

The change from summer to fall always reminds me of a comment in "The King's English" concerning the names, fall and autumn. Here it is:

"Fall is better on the merits than autumn, in every way: it is short, Saxon (like the other three season names), picturesque; it reveals its derivation to everyone who uses it, not to the scholar only, like autumn; and we (the English) once had as good a right to it as the Americans; but we have chosen to let the right lapse, and to use the word now is no better than larceny."

(Autumn is one of our borrowed words from medieval French.)

"Autumn" or "Fall," September is a Golden Month in the Pacific Northwest.

October in the Northwest

October gave a party,
The leaves by hundreds came:
The ashes, oaks and maples,
And those of every name.

George Cooper

October is part of that season when nature stages her finest production in Technicolor in the Pacific Northwest, particularly at high elevations. The big outdoor "show" starts in late September and may run into November, depending upon the weather.

If you live in this part of the United States, by all means go into the mountains in October to see the brilliant hues of maples, cottonwoods, mountain ash, aspens, larches, and huckleberries. Oregon and British Columbia offer the same flaming farewell to summer, especially in the mountains.

Those who live in or near Seattle have a wonderland of autumn colors close at hand. It's the University of Washington Arboretum where some of nature's most lavish displays are found at this time of year. The offerings include red, yellow, purple, and many in-between shades on shrubs and trees. Look for the breath-taking colors in vine maples, Japanese maples, Sargent's cherry, western dogwood, sourwood, and the enkianthus shrubs.

But we in the Seattle area don't have a monopoly on autumn brilliance. Numerous parks and arboretums in British Columbia and Oregon can equal or surpass our flames of fall; visit Stanley Park, British Columbia, or the Crater Lake Region of Oregon.

September was beautifully described in these lines by an unknown author:

Just after the death of the flowers,
And before they are buried in snow,
There comes a festival season
When Nature is all aglow.

We are rapidly approaching the time when "boughs are daily rifled by the gusty thieves, and the Book of Nature getteth short of leaves," so enjoy the Indian Summer that usually comes in October while we are "a little this side of the snow and that side of the haze."

Contrary to widespread belief, freezing weather doesn't create autumn's vivid colors. It's the sun that brings out the splashes of splendor. The change occurs because chlorophyll, the green coloring substance in plants, fades at summer's end in annuals and deciduous trees.

Other pigments, which were present all the time, become dominant, notably xanthophylls (yellow), carotenes (yellowish-orange to red), and anthocyanins (ranging from brilliant scarlet to blue, purple, and lavender).

Frost merely hastens the weakening of woody fiber called the abcission layer, behind the leaves. The cells of this layer decompose to the point where frost, wind, or a heavy rain will snap the wood, causing the leaves to drop.

Most of our trees in the wilds are evergreens, so we

don't have autumnal displays as spectacular as those in the East because there are more deciduous (winter-bare) forests there, particularly in New England. But our vine maples, native in Northern California, Oregon, Washington, and British Columbia, rank high for autumn beauty.

There are three large areas in the world where fall colors are unusually brilliant on trees. One of those regions is the East Coast of the United States and Canada. Another encompasses west-central Europe, including a portion of the British Isles. The third includes eastern China and parts of Japan.

Emily Dickinson summed up her thoughts about autumn in these words:

The morns are meeker than they were,
The nuts are getting brown;
The berry's cheek is plumper,
The rose is out of town.

Rainfall increases, as a rule, in Oregon, Washington, and British Columbia as fall fades into winter. But flowers continue to bloom in many places throughout October because killing frosts are rare before November.

Seattle's average October temperature (at the airport) is 52.2 degrees Fahrenheit, and precipitation averages 3.91 inches. Portland, Oregon usually is a shade warmer and a wee bit drier in October. Both Victoria and Vancouver, British Columbia, have October weather comparable to Seattle's. But it gets considerably colder east of the Cascade Range and in British Columbia's mountains.

November Often Balmy

In November, during the blackness of the polar night, a great mass of air lies heavy and cold over the arctic wastes. But here in the Pacific Northwest, late flowers are usually in blossom during November, and frosts haven't yet fallen heavily.

I use the word "usually" because winter can, and has, hit with awful suddenness in November. It happened in

November, 1955, when a great blanket of arctic air descended upon us early in the month. However, such an early freeze, with temperatures plunging to near zero, ordinarily wouldn't come but once in a half-century or more.

It's the mountains that help keep Western Oregon, Western Washington, and the coastal regions of British Columbia reasonably warm in November. Newcomers to the Northwest, especially those out of the Midwest and the East Coast, always marvel at the mildness of our fall and winter months and ask, "How come?"

The answer is twofold: (1) The Cascade Range in Oregon and Washington, and British Columbia's numerous peaks east of the coastal strips, serve valiantly to hold back cold air plunging southward out of Canada in the late fall and winter; and (2) the ocean has a moderating effect on our west side climate.

Seattle's mean November temperature is 44.6 degrees Fahrenheit. The mean for Aberdeen on the coast is 45.2 degrees, and for Olympia at the upper end of Puget Sound, 43.5. Portland, Oregon, is slightly warmer in November, with an average temperature of 45.3. For contrast, here are some mean temperatures for the month in Eastern Washington: Yakima, 38.4 degrees; Wenatchee, 37.7; and Spokane, 35.5.

Pendleton, Oregon, situated in the southeastern part of the Columbia Basin, lists a mean November temperature of 41.4, but the Pendleton region has been hit more than once with biting winter punches.

November is Seattle's second wettest month, with rainfall averaging 5.88 inches. Only December is wetter, and not by much. The rains are generous because of the moderating effect of the maritime air, which travels from west to east continuously. Rainfall in Portland, Oregon, follows the same pattern for the same reason.

How different our climate would be if our western mountains (the main ranges) ran east-west instead of north-south, with Portland, Seattle, Vancouver, and Victoria on the north side of those ranges. We would then have fierce winters every year.

I'm thankful for our maritime air and those guardians of our good weather, the Cascade Mountains. After all, we don't have to shovel away our November rain.

December—Dreaming of a Green Christmas

December is a gray and misty month in the Pacific Northwest, no doubt about it. Oregon, Washington, and British Columbia lie smack in the middle of the storm track which flings weather frontal systems at us for months at a time.

Is there any defense then for our cloudy and damp year-end weather? My answer is a positive, "Yes." I've seen numerous fall flowers extend their blooming season into December, and I have NOT seen many white Christmases in my lifetime, which goes back into the early years of this century.

Our December weather also brings to mind a report given to me by a neighbor. She had talked to her son in Albany, New York, one December day prior to Christmas and had listened sympathetically while he told of frozen water pipes and deep snowdrifts in his yard. She answered, "Oh, how dreadful! It's a lot different here. The gardeners at our retirement home are mowing the grass this afternoon." The son thought she was kidding.

Our snows of December are so infrequent and usually so minor that few residents have snow shovels. However, we have had some snowy Decembers, like December, 1965, when both skis and sleds were put to use on hills in and near Seattle by happy youngsters. Seven inches of snow fell in Seattle on December 23 and 24, 1965, and a trace on Christmas Day helped keep the landscape clean and white.

But white Christmases are a rarity. Seattle's mean December snowfall totals a meager 3.3 inches, and the total for Portland, Oregon, is only 1.5 inches.

Winter comes to the Northern Hemisphere on December 21. On that date the new year really begins, no matter what the calendar tells us, because each day thereafter, until June 21, is longer than the day before.

The days immediately following the winter solstice usually are colder, though. That's because arctic air continues to force its way into the United States from Canada for most of January. The sun begins its northward journey on December 21, but for most of the winter its rays are too feeble to give us much warmth for a while.

Nevertheless, Seattle's average December temperature is 40.5 degrees Fahrenheit, and Portland, Oregon's is 40.7. That's not so bad for two cities considerably farther north than Chicago, Cincinnati, New York City, and Washington, D.C.

Fair Weather

WHAT'S your conception of fair weather? There is no pat answer. Your opinions might differ from mine, and both of us might disagree with a rambunctious child who didn't like a soft spring rain because he couldn't go out to play.

Does a good day call for 100 percent sunshine? Or is a partly cloudy day acceptable? What are your temperature preferences? And what about the wind? Is shirtsleeve weather always a requirement for a fair day in May? One more question: what would you consider a fair day in the dead of winter?

I asked Norman A. Matson, formerly meteorologist-in-charge of the Seattle forecasting center, for some answers. His comments are interesting.

"We should distinguish between fair and favorable weather," Matson said. "Fair weather can be either favorable or unfavorable, or somewhere in between, depending upon the needs of people, crops, animals, birds, etc., at a particular time and place. The same applies to rainy, windy, or stormy weather."

Matson would consider a day to be "mostly fair" if the

sky was bright most of the day, even though the sky was dark and some precipitation fell for a time." Although our Pacific Northwest weather is often persistently gloomy in winter, the fair weather periods pretty much compensate for the gloom," he said.

"We usually have extended periods of fair weather in the summer, and those days are really delightful. During those fair summer days, our temperatures are moderately warm and the humidity low. Those balmy days are in sharp contrast to the hot, humid, uncomfortable weather which often occurs east of the Rocky Mountains.

"Even during the other seasons we occasionally have periods of a week or so dominated by a summer-type circulation with attendant fair weather."

Matson pointed out that outbreaks of severely cold arctic air are a rarity here, which is another climatological plus for the Puget Sound region. He also noted, "We have only a few days per year, on the average, when pollutants in the lower atmosphere are not effectively dispersed. We enjoy clean air most of the time," he said, adding that he and his family "like the weather of Puget Sound better than that in any other area where we have lived."

Harry A. Downs, who preceded Matson as Seattle's official weatherman, also chose to remain in this area following his retirement in 1965. He considers the weather of any particular day, month, or season to be "reasonably fair."

Another of my forecasting friends pointed out that a Seattleite's conception of fair weather might differ radically from the choices of an Eskimo in the Far North or a resident of Sun City, Florida. How true!

A few years ago, while vacationing in Southern California in January, I was preparing one warm, sunny day for a visit to the San Diego Zoo. An acquaintance who had planned the trip was apprehensive, however. He saw a dark cloud on the western horizon and said, "Maybe we should wait a day or two. The weather looks threatening."

I must have looked astonished, or maybe dejected, because a hasty decision was made to "brave the elements" and see the animals. The zoo parking lot, usually full or nearly so

on a sunny afternoon, had hundreds of empty spaces because hundreds of potential visitors from Southern California had stayed home, fearing foul weather. The clouds increased toward evening, but no rain fell, and I was comfortable in a sports shirt. It was good weather in my book.

An Eskimo, under a similar umbrella of sun and blue sky, might have stripped off parka and everything else to avoid heat stroke.

Most outdoor enthusiasts in the Pacific Northwest consider the weather reasonably fair if they can indulge in their hobbies without undue hardship. The steelhead fisherman in winter wears warm clothing to ward off chills and possible frostbite, and he rarely gives up, even when he has to chip the ice out of the guides on his rod before every cast. Skiers flock to the snowy slopes by the tens of thousands and are happy provided the wind doesn't blow a gale and a heavy fall of snow doesn't wipe out their view of the ski runs.

Mountaineers and other hikers ask only for weather that will give them solid footing on snow, ice, and forest trails. Joggers, bird watchers, and naturalists who frequent parks, lake shores, and quiet woodlands, pursue their hobbies in all seasons—staying home only on the stormiest of days.

One of the most delightful comments I ever received came from Henry A. Hansmeier, a retired forester, in response to my question: "What is meant by fair weather?"

I met Hansmeier decades ago on the Middle Fork of the Snoqualmie River when he was a zealous guardian of the Snoqualmie National Forest and I was an enthusiastic fisherman, hiker, and camper. He was always worried about fire in the timber and fearful that some careless person would let a campfire get out of hand or drop burning ash from a cigarette into tinder dry duff on the forest floor.

Hansmeier prefaced his remarks about weather by saying he retired on December 31, 1966, after having worked in fire control for the United States Forest Service from 1929 onward. He wrote:

"A fair weather forecast, particularly on weekends, spelled unfair weather to me because it offered more chance for fire to threaten or burn our forests, so we stayed close to the phone and radio, ready for action. Fair weather to me, on such occasions, meant a good shower. I think that weather is always fair in the long run, except for disastrous storms. Some people may consider the weather fair or good if it stays hot and dry for months. But I think of the fish in drying pools and of the birds that must range farther and farther to get a drink of water."

Hansmeier, from a forester's viewpoint, said he preferred a bit of rain on any "fair" day in summer. "A fair day can be cool, yet sunny, and offer some wind to blow away stagnant air and bring in a supply of clean, fresh air," he said.

In winter a day was fair, in Hansmeier's opinion, if the ground was covered with clean, dry snow and the temperature was down to where it would feel good to wear a warm coat. "I'd like some sunshine on such a day in winter," he said, "and it would make me feel good to hear the creaking of the snow as I walked. I'd like to see some frost on trees and shrubs, and tomorrow—still fair—could include a few flakes of snow to retouch the scenery."

Hansmeier said thunder and lightning always appealed to him, although foresters are forever afraid that electrical discharges in a thunderstorm might cause fires. He added: "Lightning, with a few brief showers, alerts all nature, and usually is followed by serenity and contentment, emphasizing the aromatic scent of wildflowers, trees, and shrubs. I can almost smell the sweet aroma of ripening huckleberries following a thunderstorm. And the air—ah!—how sweet and fresh! This kind of cleansing spells out a fair day for me."

Did you ever see Mt. Rainier at night when a full moon cast a soft glow upon Rainier's snowy slopes? I did, once, and the thrill will never be forgotten. There aren't many places in the world where a great mountain peak figures so prominently in the region's day-to-day weather. And the sight of that mountain helps to spell "fair" weather to me.

Dear Hearts and Gentle People—

Most residents simply take cool, gray summers in stride, convinced that the weather is changeable but not changing. Once I extolled our climate in print, even though the skies were temporarily overcast.

Wow! Our words of praise raised the hackles of a few folks who didn't hesitate to fire off letters to *The Post-*

Intelligencer's "Voice of the People" department. There was the Renton man, for example, who said:

"Whom does your Walter Rue think he is kidding? I have traveled throughout our 50 states, and this one has the worst climate of all from health and psychological standpoints. My company is transferring me, and am I happy!"

The ink was hardly dry on the Rentonite's comment before defenders of our weather began sending telegrams and letters. A Seattle reader expressed the sentiments of many in giving our climate a higher rating than that of Southwestern New Mexico, "the land of eternal sunshine," saying:

"The New Mexico sun is eternal all right, but there are some other things that go with eternal sunshine. Eternal dust, for one, and eternal crawling creatures . . . But what I really missed (while living there) was Seattle's marvelous air—that fresh, cool, newly washed perfume of cedars, firs and pines . . . "

We thought that letter might have silenced the dissentters, but another gray day came and, with it, a column of ours comparing our climate with that of numerous cities along our parallel of latitude. Then a woman in Tacoma took pen in hand to declare that we had made the overstatement of the year in calling our climate "superb." She asked why we hadn't compared Seattle's climate "with cities that REALLY have a good climate." And, like our Renton correspondent, she wondered whom we were trying to kid.

Once again a goodly number took time to write or phone and say: "We like it here!"

And so it goes. Some folks want it sizzling hot, with nary a cloud, while others say: "Lord, deliver us from extremes of weather and let the skies turn gray now and then, even in summer."

Someone wrote long ago that the world's best weather is found in those places "where effort and reward are so neatly balanced that mankind is stimulated but not overwhelmed by the forces of nature."

In our opinion, the Pacific Northwest qualifies (under that definition) as one of the "best places."

All the Time, Complaints!

A fellow newspaper man stripped our climate to the bare bones in one of his columns, saying it's not really the finest in the world. The only people making such a claim, he said, are those who spend the winter in Palm Springs, Scottsdale, or Hawaii.

He wasn't complaining about Seattle's wetness, although we suspect he harbors the secret hope that his next book (or the next, or the next) will bring the wherewithal to make living in the sun a possibility during the rainy season.

Good luck! And if the windfall arrives, send for me. I'll tote your golf bags, although I'll need an umbrella to fend off the sun.

Don't misunderstand. I appreciate the sun, and who doesn't? But I don't like its relentless, searing heat day after day, week after week. Perhaps the pure Scandinavian genes that determined my makeup, including a sun-sensitive skin, created a preference for repeated changes in the weather like those of Northern Europe.

One of my acquaintances, a moody man, is unhappy from November through March, the rainy season, and loudly proclaims: "This is the worst climate in the world!"

I can't agree.

I admit that Seattle gets only 49 percent of the total possible sunshine in a year's time, and that we live close to a prolific storm factory whose output of clouds and rain finds easy going through the Puget Sound Basin.

I acknowledge that some springs here are cool and moist, and that one out of every three or four summers is grayish and drizzly.

But I contend there's no better place in all the world than Seattle and environs when the blue of the sky is reflected in the shimmering waters of our lakes and Puget Sound. Our warm season is pleasant because of the complete lack of oppressive humidity that accompanies summer's heat in so much of the Midwest and East.

I like our winters, too, and would choose them unhesitatingly if I had to decide between mild, rainy weather and the cold, snowy offerings of Boreas in other northern states.

To get a proper assessment of our Puget Sound climate you have to go to oldtimers. A resident in these parts for 80 years once told me: "I have seen summers so rainy and wet that farmers on the LaConner flats couldn't cure their hay properly. I have seen snow on the level that covered my four-foot fence on the corner of Boylston Avenue and Seneca Street, and I have seen winters with roses blooming.

"I don't think there is any material change in our Seattle climate. We are all inclined to say, 'This is the worst storm, the wettest summer, the dreariest fall, the coldest winter, the damndest spring we've ever seen.'

"We magnify the unusual and may not remember all the glorious weather we've been privileged to enjoy through the years.

"There is nothing the matter with our weather, but there is a lot the matter with *us* if we complain about our Seattle climate."

I personally agree with Henry Van Dyke who wrote;

If all the skies were sunshine,
Our faces would be fain
To feel once more upon them
The cooling splash of rain.

I'll take our changeable weather any time in preference to a steady diet of sunshine.

Records and Record Breakers

PUGET Sound, one of the nation's great inland seas, and Washington's two mountain ranges—the Cascades and the Olympics—are largely responsible for the mild and salubrious weather of Seattle, Tacoma, and other communities in Western Washington. In fact, the entire Pacific Northwest lies in one of the world's best climatic zones for both humans and animals.

Following, in condensed form, are salient statistics about the Seattle area:

Latitude and Longitude

At the Federal Office Building in downtown Seattle, it is 47 degrees 36 minutes, 32 seconds North Latitude; 122 degrees 20 minutes, 12 seconds West Longitude.

Elevations

At an official datum point, 1st Avenue and James Street in downtown Seattle, the elevation is 18.79 feet. Highest measured elevations:

512 feet—just south of West Myrtle Street and slightly west of 35th Avenue S.W.
456 feet—top of Queen Anne Hill
430 feet—West Seattle Reservoir

McCAUSLAND

422 feet—Volunteer Park Reservoir
420 feet—Maple Leaf Reservoir
316 feet—Beacon Hill Reservoir

Highest Barometric Pressures

30.83 inches at 12:05 a.m., December 3, 1921
30.81 inches, January 28, 1949
30.77 inches at 4:30 a.m., January 16, 1957
30.76 inches at 3:30 a.m., November 22, 1930
30.75 inches at 10 a.m., December 4, 1959

Lowest Barometric Pressures

28.77 inches during an afternoon storm, December 4, 1951
28.80 inches at 10 p.m., January 25, 1914
28.88 inches, October 12, 1962, during the worst storm to hit Seattle in historical times.

(The lowest recorded pressure during the great 1962 storm was 28.42 inches on a ship off the Northern California coast. Washington's low was 28.54 at Destruction Island. Pressures could have been lower off the Oregon and Washington coasts because the center of the storm was at sea, but no ships were in those stormy waters at the time to record weather data.)

Highest Temperature

Seattle's all-time high reading was 100 degrees Fahrenheit recorded twice: on July 16, 1941, and again on June 9, 1955. Those readings were taken when the city's official weather data was being recorded at the Federal Office Building in downtown Seattle. The highest previous temperature was 98 degrees Fahrenheit on June 25, 1925.

Lowest Temperatures

The all-time low for downtown Seattle was 3 degrees Fahrenheit, recorded on January 31, 1893, when a government observing station was situated on the southeast corner of 1st Avenue and Yesler Way.

The all-time low at Seattle-Tacoma Airport was zero on January 31, 1950.

The earliest major freeze in Seattle came in November,

1955, when Sea-Tac recorded 8 degrees on the 15th. The low temperature at the Weather Service office downtown was 13 on the same date. That blast of cold air surged into the city on November 11, and low temperatures at night remained below freezing for eight days.

Seattle's highest tides

14.8 ft., December 15, 1977
14.6 ft., February 6, 1904, and December 5, 1967

Seattle's lowest tides

-4.7 feet, January 4, 1916, and June 20, 1951
-4.3 feet, January 15, 1949
-4.2 feet, May 31 and June 1, 1950
-4.0 feet, June 30, 1965
-3.9 feet, July 7 and 8, 1952, and May 24, 25, 26, 1955

(The gravitational pull by the moon and sun causes tidal changes, but high winds and low barometric pressures are powerful influences too. Every inch of change in barometric pressure reflects about one foot of difference in the height of the tide, so high tides in stormy weather are likely to vary considerably from predicted heights.)

Long Summer Dry Spells in Seattle

48 days: June 23 to August 9, 1922
43 days: August 1 to September 12, 1955
39 days: June 25 to August 2, 1960
37 days: July 16 to August 22, 1977
34 days: June 30 to August 2, 1958
26 days: May 5 to May 30, 1963, equaling a dry spell in May, 1928
22 days: February 28 to March 21, 1965

Long Dry Spells Early in Year

12 days: February 24 to March 7, 1949
12 days: February 20 to March 2, 1920, a leap year

The 1920 dry spell would have been longer if a little measurable rain hadn't fallen on February 19, as the days from February 7 to 18 were rain-free. There were 25 days in that February-March dry span with only 0.02 of an inch of precipitation.

Puget Sound

The Sound contains about 31 cubic miles of water, and one and a half cubic miles of that water moves with each change of the tides. The Sound covers 2,500 square miles and has 2,200 miles of shoreline.

Lake Washington

Lake Washington and Lake Union together, plus the water in the canal leading to Puget Sound through the government locks, comprise 25,000 surface acres, according to the Corps of Army Engineers.

Weather Records

Seattle's official weather records date from May 1, 1893, when a rain gauge was set up on the roof of the New York Block at Second Avenue and Cherry Street. However, the station's history shows that the taking of temperatures began on August 1, 1890, on the southeast corner of 1st Avenue and James Street.

There is no longer a downtown office of the National Weather Service. Official records for the city are now kept at the Seattle-Tacoma Airport station which opened in October, 1947. The forecasting center is in the Lake Union Building, 1700 Westlake Avenue North; it was moved from the Federal Office Building in the summer of 1972.

The Extremes—Cold

The winter of 1861-62 was one of the coldest experienced by Seattle pioneers. Below freezing weather came in mid-December and continued off-and-on until mid-March. Lake Union was ice-covered to a depth of six inches. Snow covered the ground for weeks and at one time lay two-feet deep in drifts.

January 1868—Columbia River Frozen Over

"The Columbia River is partly frozen over." Those words, written in bold Spencerian script, are found opposite the date, in an Army hospital register kept in those days at Fort Vancouver, Washington.

For the next six days the weather observer, First Sergeant E.Y. Chase, merely wrote "ditto." On the 14th he mentioned that it was "snowing and freezing," and on the 15th he reported: "The Columbia River is frozen over. Temperature 22 degrees."

Sergeant Chase must have worn longies, mittens, and greatcoat while reading the thermometer on the 16th, for there was no trace of a freeze in his exquisite penmanship as he noted a mercury reading of ten degrees, and commented: "Pedestrians crossed the river this forenoon."

The mercury continued to drop and reached a low for the month of five degrees on the 18th, which helped thicken the ice and "made sleigh riding good in the vicinity of the post," according to the sergeant.

The ice was thick enough by the 28th to bear the weight of horses and sleighs, the Sergeant said. That means of transportation, presumably between Vancouver and Portland, Oregon, continued for the rest of the month and perhaps longer, though the weather ledger only records the appearances of horses and sleighs on the last four days of January.

The ice broke on February 21, and river navigation was resumed on the 25th, the Sergeant reported.

Records of other 19th century freezes of the Columbia River are dated January and February, 1862, and January 9, 1875. The 1862 ice was substantial enough to hold horse-drawn sleighs, as well as a steady stream of pedestrians traveling back and forth between the state of Oregon and Washington Territory.

The *Vancouver Telegraph*, a newspaper, had its newsprint brought from Oregon by sleigh during that 1862 freeze.

There is no reference in the old ledgers to frost fairs, an ancient tradition in England when the Thames used to freeze—but surely there must have been wiener roasts, songfests, ice skating, and shinny-on-the-ice during those historic cold snaps, just as there were on the ice of Cranberry Marsh close to the present intersection of Empire Way South and South Alaska Street when I was a boy.

The 1875 freeze on the river began on January 9 and tied up transportation for several days, but that cold snap was of short duration.

January, 1888, was another cold month in the Portland-Vancouver area. The Willamette River, a tributary of the Columbia, was frozen solid at Portland for ten days, beginning on the 15th, but the Columbia apparently didn't ice over then, although river boats were menaced by chunks of ice.

The first major interruption of navigation on the Columbia in this century, due to freezing, occurred in January, 1909. That January was one of the six coldest Januarys in the past 62 years at Portland and Vancouver.

The entire year of 1909 was "the coolest since 1893" in the words of G.N. Salisbury, state climatologist of that era. He elaborated as follows: "A cold spell of exceptional severity and length . . . began on January 4, 1909, and lasted until the 11th. Heavy ice gorges and fields of floating ice formed in the Columbia River, causing . . . suspension of navigation."

Since 1909 the Columbia froze solid enough to disrupt shipping and support pedestrians as follows: From December 10 to 17, 1919; January 17 to February 2, 1930, and January 10 to 12, 1937.

In addition, there were spectacular ice jams on the river in January, 1922, late January and early February, 1950, and late January, 1957. Steel-hulled tugs were put to use during some of those icy winters to keep the Columbia's shipping channels open.

Prior to 1931 there were no obstructions to the natural flow of the river, but now there are eight dams with lakes behind them and four more in various stages of construction or planning. Because of these changes in the river's flow and warming of its waters (as a result of the dams), it is extremely doubtful that we shall again see the Columbia frozen in the vicinity of Vancouver.

In January, 1875, Lake Union was frozen, making it

necessary for a coal company to cut a channel so its scow could cross the lake. Some of the sluggish rivers froze too, including lower sections of the Duwamish. On January 19 cakes of ice floated into Elliott Bay from the Duwamish.

In January, 1880, the city experienced one of its heaviest snowfalls. No Weather Service observers were on hand then to leave us an official record, but we have seen references to snow "several feet deep" during that month. The barns of many farmers caved in, killing livestock. Six horses pulled a snowplow in order to clear city streets and open the road to Lake Union.

February, 1884, was a chilly month. Many small lakes froze, and ice formed on some slow-moving streams flowing to Puget Sound.

January-February, 1893. A fierce snowstorm began on January 27 and continued through February 8. Seattle's all-time low temperature downtown—3 degrees—was recorded on the 31st of January. Green Lake froze to a depth of six inches and downtown Seattle received more than two feet of snow.

Winter of 1896-97. Cool weather came early. November snowfall totaled 20.5 inches, with 13 inches of it falling in one 24-hour period. Total winter snow, 31.2 inches. The lowest March temperature in city's history was recorded on March 12, 1897, 20 degrees.

Winter of 1898-99 was severe, with considerable snow. Winter's low temperature, 12 degrees on February 3. Total snowfall for 1899, most of it falling early in year, was 25 inches.

January, 1909, was unusually cold, and 1909 as a whole was the coolest since 1893. Seattle's low temperature was 12 degrees on January 13.

January, 1913, was another snowy month here and elsewhere throughout the state. Snowfall in the lowlands of Western Washington was the heaviest since 1899. The mountains received their deepest snows in 20 years. Seattle's low temperature was 22.

February, 1910—White Death at Wellington

For three weeks, beginning on February 25, 1910, some

of the world's most destructive avalanches occurred in the snow-laden mountains of Washington, Idaho, Montana, and British Columbia. Nearly every old-timer in these parts has heard of the BIG SLIDE that swept two Great Northern Railway trains—one carrying passengers, the other mail—into a ravine in the Cascade Mountains on March 1, 1910, claiming 96 lives.

That disaster occurred at Wellington, east of Skykomish, in the Stevens Pass country.

It was wet, heavy snow that came rumbling off a hillside at Wellington at 1:43 a.m. on March 1, 1910, during a thunder and lightning storm. It is likely that the thunder jarred the "white mass of death" into movement that night as passengers on Great Northern train No. 25 slept in their coaches, and mail clerks bunked down in their quarters on train No. 27.

Railroad cars and locomotives were picked up like toys and hurled from a siding into a canyon below. Many people were imprisoned for hours. Heroic rescues were numerous.

But superhuman efforts by those who had escaped "hell's fury" couldn't save 96 of the men, women, and children who had been sleeping in the cars for days, ever fearful that death was lurking in the snow on the mountain.

A number of others might have died if they hadn't walked out previously from Wellington to Scenic Hot Springs.

Edward A. Beals, district editor, writing in the *Monthly Weather Review* said: "It was not the quantity of snow alone . . . that caused so many avalanches, but . . . the manner in which the snow fell."

Beals explained that a deep low-pressure system made its appearance off the Washington coast on February 23, 1910, and brought heavy snows to the mountains as it passed to the other side of the Rockies. He said that storm had barely sped eastward when another smashed into the Northwest and a third quickly followed. Beals reported: "The last two storms caused high winds and heavy rains that extended well up the slopes of the mountains, while at the summits the precipitation was mostly snow."

He explained that three distinct layers of snow lay in the mountains. The one on top was wet and heavy, lying over one which had less moisture. The middle layer rested on one whose surface had "almost the consistency of ice," according to Beals.

It was that slick icy surface that gave the later snows such a fast runway.

February, 1916—Seattle's Greatest Snowstorm

The snow had begun in the gloaming,
And busily all the night
Had been heaping field and highway
With a silence deep and white.

James Russell Lowell

On February 2, 1916, a front page story in *The Post-Intelligencer* said: "Caught in the grip of Boreas, Seattle experienced the worst snowstorm in years yesterday and will continue to feel its effects today . . . "

Twenty-four hours later The P-I told its readers that "beleaguered Seattle fought its relentless and insidious enemy, the snowstorm, yesterday and in the end admitted defeat." The news story went on to say:

"The white menace leveled its silent and interminable attack without cessation all day and night. The storm has played havoc with the peaceful traffickings . . . of man. Every resident is a volunteer enrolled to resist the wintry assault . . . "

More than sixty years have slipped by since that colossal storm, and nothing quite like it has occurred here since. The great snowfall was such a rarity that generations might live and die without seeing its equal.

At 5 p.m. on Monday, January 31, 1916, the downtown office of the Weather Service measured the snow depth at 6.8 inches, an accumulation of several days. Total snowfall for January, 1916, was 23.3 inches, making that month one of the snowiest in Seattle's history.

But the worst was yet to come!

On Tuesday, February 1, the weather observer once again measured the snow depth and made this notation: "Eleven inches of snow on the ground. Light, misting rain at times during the day; also some sleet. Street cars delayed on some lines; interurban train to Tacoma also delayed."

The ink was hardly dry on that 5 p.m. entry when the sky became a seething mass of gray-black clouds and the storm suddenly increased in fury. The snowfall changed from light to moderate within ten minutes and from moderate to heavy before the hour was out.

All through the night of February 1 and all day on the 2nd the snow piled up. At 5 p.m. on February 2 the snow depth at the weather station was 26 inches, and five hours later the depth was 29 inches.

Seattle's greatest fall of snow in a 24-hour period, 21.5 inches, occurred during that storm, from 5 p.m. February 1 to 5 p.m. the following day.

Many of the city's suburban areas and numerous surrounding communities had even greater depths of snow, and some of the wind-whipped snowdrifts were from four to five feet deep.

A report from Everett said three feet of snow lay on level ground. The P-I's correspondent at Anacortes said two feet of snow had fallen and the roof of the Apex Cannery had caved in.

Victoria, British Columbia, on February 2 reported its "worst day since 1862 in the recollection of oldtimers." The Victoria story said the city had four feet of snow on the ground.

Port Angeles was raked by 40-mile-an-hour winds while the storm was dropping two and a half feet of snow on the community. A newspaper man there called it the worst storm in the city's history, saying more snow (four feet) fell there in 1893, but the low temperatures and high winds made the 1916 storm more severe. He said the 1893 snow came and went quickly "without any wind to speak of."

So it went all over Western Washington and much of British Columbia. The storm began to fade on Thursday, the 3rd, but the snow lingered because temperatures were still below freezing. It took days for things to get back to normal.

The Weather Service's log for February 3 said: "Cable cars are running on James Street and Yesler Way, and street cars for short distances on other lines. Lake Union is about one-half surfaced on the north side with slush, more or less frozen into ice."

Seattle's total snowfall in February, 1916, was 35.4 inches, and the accumulation for January and February totaled 58.7 inches. Whew!

December, 1924—Green Lake's Big Freeze

I remember that freeze of December, 1924. We went to Green Lake with a group of young blades from Rainier Valley and were thrilled by the winter wonderland of flashing skates, weaving lights (held by some of the skaters), and the friendly glow of fires.

The weather records don't say, but it's our guess that skating on the lake began about the 21st after five consecutive days with minimum temperatures below 20 degrees. The low for the month was 12 degrees on the 18th.

But nighttime mercury readings were still below freezing until the 27th.

The next—and the last—major freeze on Green Lake occurred in January, 1930, one of Seattle's chilliest Januarys. By the 13th of the month ten acres of water at the south end of the lake was covered with five or more inches of ice and opened for skating with the blessings of the Park Department.

Two days later the ice had thickened to ten inches near shore and about eight and a half inches in the center, so the whole lake became a winter wonderland of gleaming blades, as well as gleaming lights after dusk. The light came from flashlights, lanterns and bonfires.

That was one of the longest skating sessions on Green Lake in recorded history—13 consecutive days.

January, 1950—The Seattle Blizzard

Seattle had a full-fledged blizzard in January, 1950. The storm claimed 13 lives and caused damage in excess of $1 million.

A blizzard is defined as a storm characterized by low temperatures and strong winds blowing a great amount of snow, mostly fine, dry snow picked up from the ground.

Under National Weather Service standards, such a storm must be characterized by winds of 35 miles per hour or higher, low temperatures, and sufficient snow in the air to reduce visibility to less than a quarter-mile.

A severe blizzard, according to the Weather Service has winds exceeding 45 miles per hour, temperatures near or below ten degrees Fahrenheit, and visibility reduced by snow to near zero.

Our storm of January, 1950, began on Friday, the 13th, when arctic air moved into Washington from British Columbia and met a vigorous mass of moist ocean air which had moved inland. Temperatures began to fall sharply. A heavy snowfall started before daybreak. The wind, which had been a gentle breeze, began picking up speed at 9 a.m.

Seattle put 19 snowplows to work that Friday in an effort to keep traffic moving. A sanding truck followed each plow, but the wind carried the sand away almost as fast as it fell.

The Weather Service's forecasting center here issued a special blizzard and livestock warning. Most schools closed early. Trains were hours late, and shipping halted on Puget Sound. Many small boats were lost. The wind picked up water from Elliott Bay and cast cloaks of ice over men lashing down vessels and cargo at the docks.

Ten inches of snow fell in Seattle on Friday. The wind reached a veloity of 14 miles per hour at 9 a.m. that day and 21 miles per hour by 5 p.m. The highest sustained velocity at the weather station on the 13th was 26 miles per hour at 9:21 p.m. However, it is believed the wind blew harder over much of Puget Sound as it whistled in from the northwest, unimpeded by hills and headlands.

November, 1955—The Big Freeze

In November of 1955 the Pacific Northwest was hit by its worst early-season freeze in recorded history. A great tongue of arctic air swept southward so swiftly that it caught every growing tree, bush, shrub, and blossom unprepared for the chill.

Flowers were blooming one minute and shriveled in death the next. Plants had no chance to harden slowly in defense against winter's onslaught.

Seattle's minimum temperature on November 10, 1955, was a pleasant 40 degrees. The next day's low was 18. That's how suddenly the cold wave hit.

How often can we expect such a freeze? The National Weather Service answers the question this way: "That cold wave was a rarity for this part of the nation at the time it occurred. We can expect such bone-chilling, crop-destroying weather in early November about once in 50 to 75 years."

You have to go back to 1896 to find a lower November average for Seattle. The mean temperature in November, 1896, was 38.4 degrees, but even in that frigid month the cold didn't come until late.

Seattle's minimum temperature during the November, 1955 freeze was 13 degrees on the 15th. The low at Seattle-Tacoma Airport the same day was six degrees.

Even Astoria, Oregon, a warm spot, had a low of 15 degrees. Washington's coastal communities, traditionally warm in November, had minimums of 11 degrees.

Seattle got through the Big Freeze with only two inches of snow. Farther south, where the first collisions occurred between Pacific and arctic air, the snows were deeper. Klamath Falls, Oregon had an eight-inch snowfall.

After the warmup, balmy weather prevailed. But the damage had been done on farms and in orchards, berry fields, arboretums, parks, nurseries, and home gardens.

Reports made to Arthur B. Langlie, then governor of Washington, indicated that damage to berry plants and vines alone in this state approximated $9.5 million, and that the strawberry industry wouldn't be able to make a complete comeback in Washington for three years.

Fruit trees were hard hit, some being killed outright and others so badly damaged by the killing of the cambium layer on trunks and limbs that they didn't recover to make satisfactory orchard specimens.

Some of the heaviest losses in Seattle occurred at the University of Washington Arboretum—countless thousands of shrubs and trees were killed. The arboretum's spring bulletin in 1956 said "The total number of rhododendrons removed or cut down to date (due to freeze damage) is well over 1,000, and that figure may be doubled eventually." The same report told of the removal of 349 dead or dying camellia plants.

The Extremes—Heat

Our Two 100-Degree Days

On July 17, 1941, *The Post-Intelligencer* told its readers: "Yesterday was the hottest day in Seattle's history. The temperature nudged to 100 degrees."

Every old-timer will remember that blistering Wednesday. Who wouldn't? One hundred-degree days are a rarity here. We've had only one other, on June 9, 1955.

The mercury climbed relentlessly on that memorable July 16th, reaching its peak at 3:37 p.m. The maximum actually was 99.9 degrees, but the National Weather Service always rounds off such figures into whole numbers. Anyway, who's going to fuss about one-tenth of a degree on a sizzling day?

Of interest, though, is a notation in the records of Earl L. Phillips, then state climatologist for the federal government, revealing that the maximum on June 9, 1955, was "100 absolute," meaning that the mercury definitely hit the 100 mark. So our second 100-degree day was warmer by the margin of a razor's edge.

Seattle's heat on July 16, 1941, shattered a record set on June 25, 1925, when the mercury climbed to 98 degrees at the downtown office of the National Weather Service.

The temperature on July 16, 1941, jumped sharply in the late morning. The reading at 10 a.m. was 77, but by 11 the mercury had risen to 91, and at noon it stood at 95, aiming pitilessly for the new record. The day's minimum temperature was 66 at 5 a.m.

Virtually everyone spent the day trying to cool off. Bathing beaches and parks were jammed, animals at the zoo were given cooling sprays of water, and many business firms closed shop during the heat of the afternoon.

Fortunately the relative humidity was low, around 30 percent, as it usually is in the Puget Sound region when the mercury climbs to dizzy heights.

The cooling of steel joints on draw bridges proved only partially effective in slowing down expansion of the steel. University Bridge was closed for 12 minutes during the noon hour so a quarter-inch could be cut off locking devices. And at 11 p.m. one span of the South Spokane Street Bridge couldn't be closed after having been opened for water traffic.

The Spokane Street problem snarled traffic for two hours while workmen were cutting off ends of old streetcar rails imbedded in the bridge's fixed roadway. The rails had expanded and intruded into the drawspan.

Seattle wasn't alone on the hot griddle. The entire Northwest sweltered, and many places in Western Washington had temperatures higher than 100. Some of the other maximums west of the Cascades during the heat wave were: Buckley, Battleground, and Quilcene, 102; Centralia, Olympia, and Shelton, 104, and City Light's Diablo Dam on the Skagit River, 106.

Eastern Washington was hotter yet. Yakima reported 107 degrees, Wenatchee, 111, and Okanogan, 112.

On the 17th, when Seattle "cooled down" to 91 degrees, a spectacular thunderstorm occurred in the evening. Some cooling rain fell, but lightning streaking out of scowling clouds caused trouble.

The 1941 heat wave persisted until Saturday, July 19, when a surge of ocean air dropped temperatures into the 70s. What a relief!

Seattle's 100-degree weather on June 9, 1955, caused the same problems that plagued the city in July, 1941, plus one other. The 1955 scorcher came so early that much tender growth in gardens and forests withered and died. The fresh tops of Douglas firs were particularly vulnerable.

The heat on that June day surged westward from Eastern Washington. The air was warm to begin with and got hotter as it flowed down the west slope of the Cascades. The wind blew at 17 miles an hour. The relative humidity was 15 percent. That wind, plus the extreme dryness, proved disastrous to new growth.

The Extremes—Floods

Winter, 1861-62, One of the Worst Floods in Northwest History

Floods came first, then bitter cold, to make the winter of 1861-62 one of the worst in recorded history in the Pacific Northwest, particularly in Oregon. There was no Weather Service in existence then to leave us detailed records of heavy rains, deep snows, and the heights to which rivers swelled. But thanks to pioneers who keep diaries, newspapermen who wrote of the havoc caused by raging streams, and temperatures included in morning reports at Fort Vancouver, Washington, we have an interesting weather legacy proving that the winter of 1861-62 was extraordinary.

The *Argus,* a newspaper published in Oregon City, Oregon, said in its issue of December 14, 1861, that rain had fallen almost continuously in November, 1861, and "a vast amount of snow must have accumulated in the mountains." There was mention of "a warm, humid, state of the air" at the end of November, indicating that a Chinook wind had sent temperatures rising in the Cascade Range and caused extensive melting of accumulated snows.

The *Argus* writer said the Willamette River, which flows past Oregon City, began rising during the first week of December. He told how "the insatiate monster" crept up, inch by inch, crushing and grinding houses and business buildings. The first light of morning on Wednesday, December 11, "revealed a scene of desolation terrible in its extent," he reported.

The newspaper said the river had carried away the Oregon City and Island Mills, the foundry and machine shop of the Willamette Iron Works, and all the breakwaters except one short piece. The heavy rainfall covered vast sections of Oregon, as well as Southwestern Washington and Northern California.

George H. Himes, who then lived in Olympia, Washington Territory, mentioned in his diary that November, 1861, was "very stormy." The rainfall at Olympia must have been heaviest on November 4, for on that day Himes wrote: "Everybody's roof leaked."

Those floods of years ago caused far more hardship than would result from comparable high water today because rivers were main arteries for transportation of most necessities of life for the pioneers, and no boats could operate on the streams for days. There was no flood warning service, either.

The rampaging Willamette River completely destroyed the town of Champoeg, where the first provisional government of Oregon was adopted. The town was never rebuilt.

Some warehouses that survived the flood burst later from the swelling and sprouting of wet wheat and other grains. When news of the flood reached The Dalles, Oregon, on the Columbia River the price of flour jumped from $7 to $12 a barrel.

Oregon's famed Umpqua River and scores of other streams flooded at the same time. A letter from Fort Umpqua, dated December 12, 1861, and published by the Portland *Oregonian* on January 9 said in part: "We have had the greatest freshet ever known in the Umpqua River. We do not know how great the damage is (upstream), but if floating houses, barns, rails, and produce . . . are any indication, the entire country has been devastated. The water has been from 10 to 15 feet higher at Scottsburg than (during) the freshet of 1853, which was higher than ever before in the memory of the oldest Indians . . . "

May, 1948, Flood Destroys Vanport, Oregon

On May 30, 1948, the sun shone brightly over Vanport, Oregon a community of 18,700 people north of Portland. Wispy clouds hung lazily overhead as if at anchor in a blue sea. The sharp staccato calls of pheasants—kook! kook! kook!—cut sharply through the still air, and robins tugged at worms in moistened lawns.

The quiet was the quiet of a Sunday morning, yet residents in virtually every house were astir. Nearly everyone, except little children, had gotten up early to watch as the mighty Columbia River (just beyond Vanport's protective dikes) hurled itself and its flotsam toward the sea. Debris from far upstream, including logs, snags, boats, fences, and lumber crashed against a railroad fill and other man-made "walls" that kept the swirling waters out of town.

Inch by inch the river rose and the water slapped more furiously at the dikes with each passing hour, but at 8 a.m. sighs of relief soughed through the city as the words went out: "You will have time to leave. Don't get excited. The dikes are safe at present."

Nearly eight hours later one of the main dikes (built in 1907) began to weave and pitch. Hundreds of men rushed to the sagging wall and began hurling sandbags, rocks, logs, and everything else at hand into the breach. Others ran through the city shouting: "Run for your lives! The dike is breaking!"

Disaster struck at exactly 4:15 p.m. The dike collapsed. A wall of water toppled onto Vanport. The city, once America's largest war-time housing project, was wiped out in less than an hour. Men and women swam frantically for the dikes. Policemen and others on dikes that remained standing threw ropes into the water and people pulled themselves along, hand-over-hand to safety. Big children grasped the ropes or helping hands and saved themselves. Little children rode on the shoulders of swimmers.

Refugees were sent to grade schools or Red Cross headquarters in nearby Portland; others were taken in by residents who lived close to Vanport, but on high ground. Oregon's National Guardmen were mobilized and sent to Vanport and elsewhere along the lower Columbia to aid refugees and help maintain dikes that still held. The Army sent 1,000 soldiers from Fort Lewis to guard property, prevent looting, and help the homeless.

The weather had been warm for days prior to the big flood, causing a heavy melt of snow in the mountains. Heavy rains during the middle and latter part of May, 1948, added to the tremendous volume of water that made its way into the Columbia.

If the Vanport dikes had collapsed in the night, the death toll might have run into the thousands. Those who died—perhaps as few as 20—were trapped in their homes by rising water, or swept away before one or another of the human chains could reach them. Damage at Vanport and elsewhere along the swollen Columbia amounted to $100 million, according to estimates. The flood, one of the worst in the history of the West, lasted 45 days.

But Vanport "died" in an hour and never came back to life.

That 1934 Blow

One of the worst storms in the city of Seattle's history blew up from the southwest on Sunday morning, October 21, 1934, and the western sections of both Oregon and Washington bowed before its fury.

The wind reached a velocity of 70 miles an hour in Seattle and 90 miles or more per hour in other areas. Some frail homes were blown to bits. Roofs came off several industrial plants. Part of a small hotel in Seattle's South End collapsed, killing a man who was buried under tons of brick and other debris.

Fishing boats and pleasure craft were tossed about like kindling wood on the water. Five Seattle fishermen drowned when the purse seiner, *Agnes,* foundered in furious seas near Port Townsend. A Seattle newspaper man, Nicholas Schwartz, was pitched into Silver Lake, near Bellingham, from a canoe and drowned.

The most spectacular wind damage occurred in Seattle's harbor, where the trans-Pacific liner, *President Madison* tore loose from her moorings and rammed and sank the steamboat, *Harvester.* The *Madison* also hit two other ships and smashed into a ramp on a dock before coming to rest in a snarl of wreckage around her.

The smokestack of the central heating plant at the Church of the Immaculate in Seattle toppled and crashed through the dome of the sanctuary. Parishioners had left only ten minutes before.

A hangar at Boeing Field was picked up by the wind, then dropped onto four planes it housed.

So it went. Death and destruction mounted by the hour. At the end of the Big Blow, 18 persons were dead in the Northwest, damage ran into millions, and scores of communities were without power or phone service. Restoration of light, power, and phone lines alone took days.

The Columbus Day, 1962, Storm

Little did anyone suspect that Freda, a lazy little typhoon born near the Caroline Islands, east of the Philippines, was going to take dead aim for the United States and become one of the most destructive storms in American history.

Freda was just a restless whirl over the Pacific Ocean when first spotted by aircraft. She meandered aimlessly at first while picking up moisture from the warm seas.

The energy transferred from ocean to cyclone gave Freda muscle enough to become a full-fledged typhoon (another name for a hurricane), with winds traveling counterclockwise at 73 miles or more per hour.

Her position on October 3, 1962, was fixed by spotters at 30 degrees North Latitude and 159 degrees East Longitude, between the Mariana Islands and Wake Island. She was churning the ocean but causing no damage, as ships kept out of her path and American aircraft kept a respectful distance while watching.

Between October 3 and 4 Freda swung east to the 165th meridian, East Longitude, then curved to the northwest as if intending to menace Japan. She had the tropical typhoon eye around which the winds blew furiously, but the whole cyclonic system moved slowly until October 8, when Freda showed signs of becoming an oddball among the great storms of this century.

On October 8, while at 29:06 degrees North Latitude and 164:02 degrees East Longitude, Freda began showing signs of fading out as a typhoon and becoming an extratropical cyclone, typical of the whole gales that hurl themselves at the American continent from the North Pacific Ocean. She was then about 1,750 miles southwest of Tokyo.

Freda was 5,500 miles away from the closest North

American land then, and she seemed destined to blow herself out over water.

But strange things were happening to the typhoon. It started to lose its eye, and it picked up speed, heading northeast toward the Aleutian Islands.

Between the 8th and the 10th Freda zipped from the 30th to a point near the 47th parallel of latitude in a long arc, a distance of 2,800 miles. By the 10th all typhoon characteristics were gone.

After crossing the 175th meridian, West Longitude, Freda changed course again. She was then about 2,500 miles due west of Seattle, and moving more swiftly with each passing hour. But her new target seemed to be Mexico, not the United States.

On the 11th, the monstrous "low" had a well-developed front. Between 4 a.m. and 4 p.m. that day it moved across 1,400 miles of ocean.

Where was Freda headed? No one knew, and not until Columbus Day, the 12th, did she reveal her ultimate goal, the West Coast of the United States. On that day, while approximately 1,200 miles due west of Los Angeles, she turned sharply once more and headed for the Pacific Northwest, where she arrived with awesome fury in the afternoon.

Freda's center remained at sea but her diameter was so great that her winds lashed hundreds of miles of ocean and land at the same time. On the eastern side of the storm the winds were from southerly quadrants, and their velocity was intensified by the storm's own rapid rush northward.

We have referred to this storm of Columbus Day, 1962, as Freda simply for convenience, but it was no longer a typhoon, hence no longer entitled to a name. Yet it probably was more violent on the 12th than at any time after its birth as a true tropical cyclone.

The storm tore its way through forests, residential areas, and industrial sites in northern California, Oregon, Washington, and British Columbia before whirling itself to death in Canada's mountains. It scoured a path of death and ruin for 1,000 miles.

It struck in the same devastating manner as the famous

New England hurricane of September 22, 1938, and equalled or eclipsed that storm for sheer savageness and extent of damage.

The Columbus Day storm claimed 46 lives, blew down something like 15 billion board feet of timber, damaged 53,000 homes, and destroyed thousands of utility poles and hundreds of miles of utility lines.

Red Cross records show that 317 persons received injuries serious enough to put them in hospitals, and the number of families who suffered losses of one kind or another numbered more than 75,000.

The value of the timber blown down was $750 million, much of it salvageable. But the cost of building roads into many devastated areas was staggering. Future losses were in store too, from fire and infestation of the forests by bark beetles.

Beetles and certain other insects multiply quickly in blown-down timber and can cause incalculable damage, as can fire that roars through storm debris.

The West Coast actually had three intense storms between October 11 and 13, 1962, but the granddaddy of the trio was the ex-typhoon.

Storm damage in Oregon was estimated in excess of $170 million. Washington's toll was around $50 million, British Columbia's $10 million, and California's $5 million.

Although Oregon was hardest hit, the Naselle radar site in Pacific County, Washington, reported the greatest wind gust, 160 miles per hour, and a sustained wind speed of 150 miles per hour for a time.

Other high wind speeds in Washington were 100 miles per hour at Renton, 92 miles per hour at Bellingham and Vancouver, 88 miles per hour in Tacoma and 83 miles per hour at West Point, Seattle.

Oregon's strongest winds in gusts were 131 miles per hour at Mt. Hebo's Air Force station, 116 miles per hour at the Morrison Street Bridge in Portland, and 106 miles per hour at Troutdale.

Those of you who were here will remember that power failures occurred all over the city, trees snapped, signs blew down, log booms broke up under the furious pounding of the waves on Puget Sound, and hundreds—or was it thousands?—of small boats broke loose from their moorings.

Even the Washington State ferries ceased operating at the height of the storm. And the grounds of the Seattle World's Fair were closed early as a precautionary measure.

Portland's Multnomah Stadium had part of its roof peeled off by the wind, and there was so much debris everywhere that it was feared the Washington-Oregon State football game the following day might have to be postponed. But a mammoth cleanup job went on through the night and into the morning making it possible for the game to be played. Many fans who intended to drive to Portland for the game couldn't get there, however, because of road closures due to fallen trees.

The lowest known pressure in the West during the storm was 28.42 inches, recorded on a ship off the northern California coast. Washington's low was 28.54 on Destruction Island off the coast.

Storms of such magnitude are rare. We might not have another one as fierce for 25 years or more. And that would be too soon.

Washington's Tornadoes

A funnel cloud snake-danced September 11, 1966, over South Seattle—a swirling twister large enough, weather observers said, to have caused damage had it hit the ground.

E. Lee Whitmire, National Weather Service observer on duty at Seattle-Tacoma Airport, said he saw the funnel about three miles north of the airport at 2:06 p.m. and watched it for five minutes until another cloud blocked his view. He said the incipient tornado was tracked on radar for 15 or 20 minutes before it dissipated in the foothills of the Cascades. It moved east-southeast, he said.

Whitmire estimated the funnel cloud was about 11,000 feet long and that its top was at an altitude of about 12,000 feet. "The closest it got to the ground was about 300 feet," Whitmire said. "It was quite extensive. From my line of sight, the top would have been about 40 degrees. It must have been 50 or 60 feet wide at the bottom. If it had hit, it would have done some damage. We were very fortunate. I used to live in tornado country, in Illinois."

He said the funnel was "pure white, like a waterspout." He explained the funnel would have been filled with debris if the twister had touched the ground. "It possibly was the remains of a waterspout that started over Puget Sound and then went back up into the clouds before it hit land," Whitmire theorized.

A similar miniature tornado was reported August 18, 1964, east of Kent, between Lake Meridian and Lake Youngs. That tornado's funnel also swayed like a wriggly snake. It remained in the air most of the time, but touched down several times in the forest-covered hills.

A Weather Service technician living near Lake Meridian saw the 1964 tornado twist and turn toward Issaquah and noticed debris in it, although he said its funnel was "mostly white."

Tornadoes are extremely rare in the Pacific Northwest but not unheard of. Seattle's first confirmed tornado gouged a furrow of destruction between View Ridge and the Sand Point Naval Air Station on September 28, 1962. That funnel shattered windows, uprooted trees, splintered heavy timbers and, in one instance, hurled a carport over the roof of a house. The 1962 tornado crossed Lake Washington in the form of a waterspout, then cut a swath through trees between Juanita Point and Denny Park on the East Side before disappearing.

On December 12, 1969, a tornado played hit and run with South King County ripping roofs, jostling cars, and toppling trees and poles before spiraling eastward into the Cascades. No one was injured. Some were scared.

Damage was substantial when home and business owners totaled up the amount of lost or damaged roofs and splintered siding.

Some accounts had three separate twisters. But two were weak and did a quick fadeout. A third, officially recognized by the National Weather Service, cut an angry swath from southwest to northeast King County.

Heavy, gusty winds and a biting thunderstorm with showers hit near Tacoma, on Vashon Island, at Des Moines, Midway, and Kent and swooped across Interstate 5 south of Seattle, upending a small camper.

A few motorists, eyeing the darkened skies, bumped into each other. Some got out for a better look.

One motorist said the tornado appeared to rise from the Puget Sound Basin.

Its menacing funnel could be seen racing toward Interstate 5. He said: "Then the funnel lifted skyward and formed a cloud as it neared the Freeway. It reformed and touched down west of the Freeway."

The driver stopped his car and watched the twister upend a camper and strew its contents along the busy Freeway. It pummeled one small car across two lanes. "People got out of their cars in the heavy wind and watched. The tornado headed toward Issaquah and sort of disappeared."

The sheriff said roofs were ripped off homes on Vashon Island, Des Moines, and on Russell Road south of the O'Brien Bridge near Kent. Reports of popping windows came in from a variety of locations. Tornadoes create vacuums along the way and cause windows to pop outward from their frames.

Hail—a frequent companion of a twister—fell in many places, briefly. The Midway Furniture Company lost most of its roof. And a Midway drive-in reported a small car blown over and against a truck.

Highline Community College said about ten parked cars were damaged by trees toppled by the tornado and windows in some other cars popped.

Signs took a beating. Small ones went sailing or crashed to the ground. One Highway 99 large-sized billboard caught a gust head-on and bent groundward without snapping its metal fastenings.

At Kent, Boeing Company employes reported seeing three funnels. Two quickly disappeared, they said, but the third spun across office and laboratory buildings, ripping roofing and gouging siding.

Controllers at Seattle-Tacoma International Airport and Boeing Field were warned, but the twister hit no planes or towers.

The Weather Service declared an "all-clear" by late afternoon but kept up small-craft warnings.

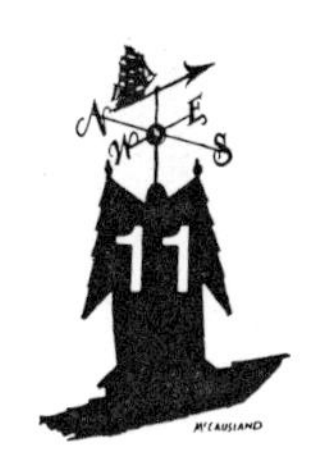

History

LONG before the National Weather Service was established, the nation's Army surgeons were keeping weather "diaries." One of the purposes was to study the effects of various climates upon the health of the soldiers, but the Army also was cognizant of the need for reports on the types of weather it might have to contend with in fighting on the American continent.

I came across two of those old "diaries" at the Weather Service office in Seattle and found them exceedingly interesting. One contains reports on weather observations taken at an Army hospital on Cape Disappointment, Washington Territory, from June 1, 1864, to April 14, 1871; the other gives descriptions of the weather at Vancouver Barracks, Washington Territory, between July 1, 1866 and June 30, 1880.

The surgeons usually made terse reports, such as these at Cape Disappointment (near the mouth of the Columbia River):

March 13, 1871—"Stormy and ugly weather."
April 6, 1871—"Fearful squalls."
April 7, 1871—"Tremendous rain and hail, full of alarming

consequences." (The alarming consequences weren't explained.)

Regular observations of temperature, wind, and general state of the weather ceased May 29, 1869, at Cape Disappointment, but someone resumed the work in October, 1870, and the keeping of the diary continued spasmodically until April 14, 1871.

The nation's first country-wide weather system was inaugurated in 1814 by Dr. James Tilton, who was then surgeon general of the Army. Dr. Tilton ordered all surgeons attached to regiments, and all post and hospital surgeons, to keep "diaries of the weather."

The war then in progress prevented perfection of the system, so another order concerning weather diaries went out in 1818 from the office of Dr. Joseph Lovell, successor to Dr. Tilton as surgeon general.

Dr. Lovell asked each Army surgeon to keep a weather diary and to note "everything of importance relating to the medical topography of his station." He particularly wanted reports on the climate at each Army post and studies made of illnesses prevalent in the vicinity.

Both Dr. Lovell and his predecessor were concerned about "complaints" endemic in various parts of the country. Dr. Lovell said he also wanted to know, "whether in a series of years there be any material change in the climate of a given district of the country; and if so, how far it depends upon cultivation of the soil, density of population, etc."

The Army saw an excellent opportunity for such studies, not only in heavily populated areas, but also on the ever-changing frontiers. Scientists helped the surgeons improve and broaden the weather reporting system. Prior to 1843, the thermometer was the only instrument in use, but barometers and other instruments were distributed in that year.

An early history says that the task of taking weather data was "sometimes onerous," yet the surgeons faithfully kept their diaries, although burdened with caring for the sick and wounded.

The armies carried meteorological instruments even while engaged in wars with the Mexicans and with Indians.

After the establishment of a weather service by the United States Signal Office in 1870, the meteorological work of the surgeon general's office was discontinued. But many Army surgeons continued weather observations which were "gratefully received by the Signal Office."

The National Weather Service has been in charge of nationwide reporting systems since 1890.

1870—Army Signal Service Became Weather Service

"Each weather observer in charge of a station will make arrangements . . . for some competent person to perform his duties in the event of sickness or disability . . .

"At all stations where there is a private soldier on duty as assistant, the observer will instruct the private in the theory as well as the practice of meteorology, signaling and telegraphy . . .

"Observer sergeants will wear their uniform coats buttoned and be neat and careful in their dress. Any negligence . . . will be considered sufficient cause for the reduction of the offender to the rank of private soldier."

Those directives are from one of the most interesting ledgers I ever saw, the annual report of the Chief Signal Officer, United States Army, in 1872.

The Signal Service had taken over operation of the nation's weather network in 1870, and it pursued a no-nonsense policy from the start, just as the National Weather Service does today.

Thanks to the state climatologist for the Weather Service, and keeper of many old records, I was privileged to dip into the history of Army life (and weather observing).

The Signal Service detachment of the Army had taken over responsibility for weather observing from Army surgeons, and in 1872 was deeply involved in creating an extensive network of weather stations.

The 1872 report emphasized that observer sergeants

must make their reports "absolutely correct," and any shortcomings in that respect would "render them liable to punishment." It was stressed that "a single incorrect report might cause loss of life and property . . . and all reports must be made with this responsibility clearly in view."

Economy was a watchword, too, and each observer sergeant was advised that he could rent one room for the performance of his duties (if the station was not at an Army post), but under no circumstances should he contract for rental of the room for more than $18 per month without special authorization.

The instructions added that: "If any article of public property be lost or damaged through the neglect or fault of any observer or assistant, the money value thereof will be stopped against his monthly pay."

The report said a one-room weather station could have one desk (price not to exceed $10); one table, with drawers (stained pine and not to exceed $6); one wash-stand (at a price not exceeding $1); four chairs (not to exceed $2.25 each in cost); one stove with pipe (not to cost more than $25), and such necessities as a fire shovel, water bucket, cup or dipper (but not both), broom, brush, dustpan, goblet, lamp, lantern, oil can, and two spittoons.

The Signal Officer's report included weather information from many parts of the country, but I could find nothing for Washington Territory.

Portland, Oregon, was included though, and the rainfall there from October 1, 1871, to September 30, 1872 (12 months) was given as 42.32 inches. It was a wet year in the Northwest in 1872!

1886—Joshua Green, Sr., Came to Seattle

One chapter in the Pacific Northwest's climatological history came from the warm heart and keen mind of a beloved pioneer, Joshua Green, Sr., several years before his death in 1975 at the age of 105.

Green saw nearly 90 years of our weather come and go, and he credited the climate here (in part) for his longevity. I asked the venerable dean of Seattle's bankers two questions: Is our climate changing, and what do you like about this part of the United States?

He got right to the point, saying: "When I leave my office or home, to enjoy a little temporary sunshine after a strong southwest wind and rain, the filtered air is so sweet, soft, and fresh that you can actually taste it. And p.d.q. the rain has run out of sight down the hills, and there is no stagnant water left where bugs and germs can breed. And you don't have to boil the water but can drink it right out of the faucet or the streams."

Green said he was always thrilled to come back to Seattle after his hunting trips in equatorial Africa and other hot places. He said the Northwest, on such occasions, made him think of the Garden of Eden "or even Paradise itself."

He was of the opinion that the climate of much of the Northwest is getting warmer, but he didn't think Seattle, itself, is warming up. He said this area's winds and rains from the southwest and southeast have not changed perceptively since he arrived here in 1886 from the Deep South.

Green recalled that he had seen only three or four severe winters with deep snows. Many a winter has come and gone, he said, with temperatures so mild that flowers continued to bloom in December and January. He noted that "the warm and cold years seem to come in cycles, but the warm ones are the most numerous." He added:

"I saw one winter here with snow almost three feet deep. The cold snap was followed by a vigorous chinook wind which warmed us quickly, and the snow was gone in three or four days. Those chinooks are healthy for me. I can sit in a duck blind all day with a chinook wind blowing and a warm rain falling, and suffer no ill effects. But one must be dressed properly. You can't let the water run down your back (from a carelessly placed collar) into your hip boots and down to your feet. That will give you cold feet, which is bad indeed, as the courts construe 'cold feet' as incompatibility of temper, which is grounds for divorce."

(The chinook wind, which Green liked, is a wind coming into the Northwest from the warm ocean to the southwest, often from the vicinity of the Hawaiian Islands.

Another kind of chinook is a warm, dry flow of air down the east slope of the Rocky Mountains. The air in the mountain chinook warms itself by compression as it sinks into the lowlands. Eastern Washington occasionally gets a chinook of that type which comes down from the high Cascades.)

Green said he had been shooting ducks in the Puget Sound country for most of his life "with many men of many minds" and hadn't yet caught pneumonia. Nor had any of his hunting partners suffered weather-caused illnesses, he reported.

1893—First Entry in Seattle's Weather Ledger

The first entry in Seattle's first official weather ledger is dated May 1, 1893. It relates that G. H. Willson, observer, United States Weather Bureau, arrived in the city on February 27 and immediately began a round of calls upon officials of the Chamber of Commerce, the Merchants and Manufacturers' Exchange, the newspapers, the Board of Health, the University," and other interested in the work of the Bureau." (The Weather Service formerly was known as the Weather Bureau.)

It was Willson's job to determine "the needs and wishes of the public" and to set up a weather station. On March 22, 1893, according to Willson's notations in the ledger, authority was received to rent the rooms offered by the New York Block (the predecessor of the Dexter-Horton Building at Second Avenue and Cherry Street) at $30 per month

The station was equipped during April, 1893, and on May 1 the first observation was taken. The entry for May 1 says it was cloudy but a pleasant day with light to fresh southerly winds.

Other interesting entries:

July 10, 1893—Partly cloudy but pleasant. The clock stopped and could not be made to run. A comet was observed by several persons. Anemometer oiled.

August 25, 1893—Dense fog in the morning, clearing . . . about 11 . . . then pleasant. Some parties reported feeling a slight earthquake this morning, some giving the time of the shock as 8:30, and others, 8:25.

August 31, 1893—Dense smoke, with light variable winds. A hot wave passed over the city between 5:30 p.m. and 8:30 p.m., reaching its maximum, 90.8 degrees about 7 p.m.

September 1, 1893—Dense smoke all day. The sun appeared like a ball of fire and was not brilliant enough to hurt the naked eye after a long view.

Entries for September 2, 3, 4, 5, and 6 of the same month also mentioned dense smoke. The source of the smoke was not given, but it is probable that forests were ablaze in parts of the state. Great palls of smoke often hung over the Pacific Northwest for weeks in those olden times as the result of forest fires.

A welcome rain that cleared the air of smoke is mentioned under the date of September 8, 1893. The observer also wrote: "Received annual supplies. Clock regulator very badly damaged en route; careless packing."

The entry for November 29, 1893, says the day was "the darkest of the year," with smoke, fog, clouds, and an "Oregon mist" making it necessary to use lights in office buildings.

A severe storm is described under the date of January 15, 1894. The wind reached a velocity of 37 miles an hour, the observer said. He attached a clipping from *The Post-Intelligencer* of January 16 which reported that the storm did much damage in the harbor and on adjacent waters. The clipping told of telegraph lines being "prostrated in every direction" and said three ships dragged their anchors for long distances and two of them nearly crashed into a pier near the foot of Seneca Street.

On February 16, 1894, the Weather Bureau gave the Seattle office permission to hire C. E. Hartman of Buckley as a temporary assistant due to "the illness of Mr. Pancoast." It was noted that Hartman would receive $1.50 per day.

The first 19 days of February, 1894, were mostly cloudy and wet, but on February 20 "the weather changed . . . and at about 9 a.m. the sky was as pure a blue as could be seen anywhere," the observer wrote. He added: "This has been a glorious day with light northerly airs and beautiful sunshine. The amount of snow on the ground at 8 p.m. was about one inch and (is) melting rapidly. Mr. Pancoast is still sick."

An account of a brilliant auroral display is given on February 22, 1894. A newspaper clipping, which was made part of the official entry, said the aurora consisted of two white rays of light which rose from the northern horizon, shooting upward toward the zenith, lighting up the whole heavens. They "looked like sunbeams in a dark cellar," the writer said.

A summary entered in the ledger on March 31, 1894, said "this has been the wettest March since 1890," but it also said "the crop season is much more advanced than last season; fruit trees are in bloom."

The reference to earlier years indicates that cooperative weather observers had taken temperatures and measured rainfall here before the Weather Bureau's office was established.

A violent storm is described in the entry for April 25, 1894. The ledger says "a sudden and terrific gust of wind" at 9:55 p.m. blew the wind vane support down, breaking it in two places and wrecking the vane. A clipping from the *Seattle Telegraph* of April 26 said the sloop *Oseo* from Whatcom with three men aboard overturned near Duwamish Head, but the *Oseo* "lay well up in the water" and the trio clung to her until rescued.

Questions and Answers

WHEN I was writing a weather column for *The Seattle Post-Intelligencer*, readers asked hundreds of questions about wind, rain, snow, ice, temperature, humidity, and other facets of meteorology.

Some of those questions and answers belong in this book because a new generation has come along, wondering about the same things. So here we go:

Q. I once saw the waters of Lake Washington furiously agitated in a line stretching from Mercer Island to Seattle, but the width of the "tempest" was very limited. What happened?

A. I witnessed a similar phenomenon during an early-day Gold Cup race. The fury was caused by the clashing of a north wind and a south wind. That happens occasionally when a swift west wind is diverted by the Olympic Mountains, with one part of the air stream going for an end-run around the north slopes of the mountain range, and another part going around the southern hills. The two air streams sometimes converge on the east side of the Olympics, and when they meet they clash violently. Such "tempests" are most apparent on the water.

McCAUSLAND

Q. What causes the thunder during a lightning storm?

A. The air "explodes" with a bang along the lightning's path.

Q. What is meant by the term, "black frost?"

A. It's an Old World name for a cold wave unaccompanied by white hoarfrost. Ordinary frost (hoarfrost) forms when moisture in the air condenses at freezing temperatures.

Q. I was born on December 12, 1919, at Port Angeles and have been told it was a chilly day. How cold?

A. The low temperature was 13 degrees Fahrenheit.

Q. Has Washington ever had a true hurricane?

A. No, but we've had winds of hurricane strength. There's a difference. A tropical storm, born near the equator over warm water, becomes a hurricane when winds reach 74 miles per hour, but it has cyclonic characteristics all its own, including a central eye where a calm prevails. The wild winds that smash into the Pacific Northwest are a different breed of storm which originates over the North Pacific Ocean.

Q. What is meant by "whistling for the wind?"

A. The expression goes back to sailing ship days when vessels occasionally encountered dead calms. Old-timers on becalmed ships told apprehensive youngsters that, "You can whistle for the wind if you really want it." Many gullible boys believed the jokesters and often did their whistling just before the welcome breezes came, causing them to think they had dispelled the calm.

Q. I've heard of fish, frogs, earthworms, and other strange things falling from the sky. How come?

A. The wind carries all sorts of things, living and dead, into the atmosphere. Winds rising vertically out of tornadoes, hurricanes, waterspouts, and dust devils can easily suck up earth, water, leaves, and other things normally earthbound. A tornado, for example, might pick up all the fish in a pond and perhaps all the water too. But eventually, what goes up must come down.

Q. Why does the rain leave spots on my car?

A. Because each raindrop forms around a nucleus of

dust, salt, or some other minute object in the air. The water hits your car and drains away, but the nucleus of the raindrop may remain.

Q. What is wind chill?

A. Freezing temperatures, accompanied by wind, can be dangerous because the wind adds to the chilling effect. Human skin would be affected in about the same way under the following conditions: temperature 40 degrees Fahrenheit and wind speed 13 miles per hour; temperature 30 degrees and wind speed six miles an hour; temperature 20 degrees and wind speed slightly less than three miles per hour; temperature ten and wind speed less than two miles per hour; temperature zero and wind speed one mile an hour. Those are examples of the wind chill effect. So beware of low temperatures when the wind is blowing!

Q. How much ice is there in Greenland?

A. The ice covers about 700,000 square miles of the island, leaving only small areas where Greenlanders live. The ice cap engulfs about nine-tenths of the country and reaches an elevation of 11,190 feet at one point. Greenland is still firmly in the grip of the Ice Age.

Q. What is the normal decrease of temperature with elevation?

A. In dry, stable air the mean drop is 5.5 degrees Fahrenheit per 1,000 feet. But don't try to apply that exact figure to the lowering of winter temperatures in Washington's mountains. We get a variety of air masses here, some warm, some cold, carrying varying amounts of moisture.

Q. Did a warm December follow the Puget Sound area's chilly November in 1955?

A. Yes. On Christmas Day, for example, the high temperature at the downtown office of the National Weather Service (an office now abandoned) was 53 and the low only 47. The month's maximum reading was 58 on the 11th and the minimum 28 degrees on the 18th.

Q. I know that some seas are saltier than others, but what's the average amount of salt in the oceans?

A. About 35 pounds per 120 gallons of water.

Q. When did the *S.S. Islander* sink in Alaska? Did she founder in a storm?

A. The Islander went down in Stephens Passage, Southeastern Alaska on August 15, 1901. She had struck an iceberg.

Q. I believe I once saw a faint rainbow in the moonlight one autumn night. Were my eyes deceiving me?

A. You probably saw a rare moonbow. I've heard of the phenomenon but have never seen one.

Q. What is the difference between evening and night?

A. People in the Northwest generally refer to the hours from 6 p.m. to 9 p.m. as evening. Night would embrace all the hours of darkness.

Q. How much snow falls at the upper end of Lake Chelan in winter?

A. About 123 inches.

Q. I always see the rainbow as part of a circle. Does it ever appear as a full circle?

A. At ground level we see only the upper part of the circle because the source of light (the sun) is above the observer. Travelers on high-flying airplanes have seen the full circle when the sun was near the horizon and other conditions were right.

Q. Do the records prove that a bad year "comes in swimming?"

A. No. The best weather in Seattle's history was recorded in 1958, and the year started "dripping wet."

Q. Why can't the National Weather Service tell us exactly when a storm will arrive? I've known occasions when they missed by many hours.

A. Because storms travel at varying speeds. Some are poky, others move swiftly. And, at times, a storm will suddenly change direction and completely miss a region it first appeared likely to hit. High pressure areas act as buffers to hold back many low pressure cyclones. But the Weather Service does a commendable job, nevertheless, with the aid of satellite pictures of weather fronts.

Q. Is a waterspout a tornado?

A. If a funnel cloud touches the ocean or a lake and sucks up water, a waterspout is born. It's exactly the same kind of storm as a tornado over land.

Q. Which year was the warmest in Seattle, 1934 or

1958? You have referred to 1958 as the "perfect year," but my grandpa thinks 1934 was just as balmy.

A. In Seattle, 1958 was warmer by exactly one degree. The 1958 average temperature was 56.2 degrees Fahrenheit; the average for 1934 was 55.2. But surely 1934 was a salubrious year with a lot of sunshine throughout Washington and adjacent states. No year prior to 1934 was as warm in the history of weather observing in this state.

Q. Weather forecasters often refer to gusts of wind, but no one ever explains why gusts occur. Why?

A. Not all parts of the flowing air mass move at the same speed. Wind movements are comparable to ocean waves. You may see a big wave, followed by a succession of smaller ones. A wind gust is like a burst of water past a barrier within the main current of a river. Topographical irregularities add to the friction encountered by surface winds. The wind is twisted and turned by hills, trees, buildings, and other obstacles. No wonder its speed varies.

Q. Is the humidity high or low in caves where cheese is cured?

A. Mostly high. Bleu cheese, for example—and some of the other Roquefort types—are cured under mold-producing conditions. Bleu is usually dry-salted for a week or more in a room where the temperature hovers between 46 and 48 degrees Fahrenheit and the relative humidity is around 95 percent. Further curing goes on for months in moist rooms or caves.

Q. How big is the average high pressure system in the atmosphere? Are highs bigger than the lows?

A. There is no such thing as an average high or low. Some are huge, some small. The Pacific high, which expands in the summer, sometimes covers nearly all of the North Pacific Ocean (east side), plus great chunks of land in the Far West.

Some of the wintertime lows are equally huge. A low pressure system on February 23, 1957, embraced nearly nine million square miles of water and land. That low was centered over the Pacific Ocean but also controlled the weather over most of the western third of the United States and Canada.

Q. Has hail ever killed human beings in this country?

A. Yes. On May 13, 1874, "several people and a great number of sheep, geese, and the feathered inhabitants of the woods" were killed at Winnsborough, South Carolina, when chunks of ice, some nine inches in diameter, fell from the clouds. The ice chunks were formed by the coalescence of many hailstones.

Q. Why doesn't our Northwest usually have hurricanes and tornadoes?

A. It takes violent clashes of different types of air masses to create tornado weather. We rarely get such "collisions" here. A few funnel clouds have formed in this region but didn't grow into monstrous whirls. The hurricane is a storm that develops over warm water, close to the equator. Some hurricanes have pushed into northern states, east of the Rocky Mountains, but a hurricane is really in its death throes over land. The energy is derived from the warm ocean waters. Both tornadoes and hurricanes find the going easiest over flat terrain. Mountains will tear them to shreds.

Q. How did "weather" and "rain" get their names?

A. They are nouns from the Old World. Rain was spelled "regn" in Old English. The spelling is only slightly different in northern European lands.

Q. How does a thermometer tell the right temperature?

A. The mercury in a thermometer goes up when it's warm and down when it's cold (expanding in the heat and contracting in the cold). The movement is through a bore in the glass. The maker of a thermometer discovers, through trial and error, just how big to make the bore, and how much mercury to put in the bulb so it will register the correct temperature or come close. After making those determinations, the manufacturer can produce thousands of instruments with the same-sized bore and the same amount of mercury.

Q. What causes water to evaporate?

A. Dry air soaks up water quickly. If the weather is sunny and warm, the air becomes "hungry" for water. So if water is available in a pan, a pond, a lake, river, or ocean, the dry air soaks it up. Condensation, the opposite of evaporation, occurs when the air becomes saturated with water, meaning it is full to overflowing. The excess water condenses. It may form only a cloud thousands of feet above the earth or fog close to the ground. But if there is a great amount of moisture, it may fall as rain, hail, or snow.

Q. Why do we sometimes see clouds in certain places, but not elsewhere?

A. Scattered clouds develop when only part of the air has reached the condensation level. The air between the scattered clouds is too dry for condensation to occur.

Q. Why does the wind blow?

A. Wind is air in motion. The air moves in vast counterclockwise whirls in storm systems, sometimes reaching hurricane force. Air also tends to flow like a river from an area of high pressure to a place or places where the pressure is lower. Bands of air are always moving somewhere in our atmosphere. Sometimes (especially in high-pressure systems) the wind is just a gentle breeze. What a sad and foul-smelling world it would be if we didn't have winds.

Q. How low has the temperature dropped in Greenland?

A. The all-time low was 86.8 degrees below zero, measured on January 9, 1954, at an ice station operated by a British expedition. That was at 78 degrees 4 minutes North Latitude and 38 degrees 29 minutes West Longitude.

Q. I'm having an argument about Seattle's weather in the spring of 1943. I contend it was cool. I remember because my garden got a late start. What information do you have?

A. You are mostly right. March temperatures were close to normal (slightly under), but it was a rainy month. April was warm, May cool.

Q. What is the dewpoint?

A. It's the temperature, at a given pressure, to which the air must be cooled to become saturated. You could also say it's the point to which the temperature must fall to form dew.

Q. How much concentrated water does snow contain?

A. The water content isn't always the same. If the snow is "wet," it would melt down to about an inch of liquid for every ten inches of snow. When the snow is "dry," the ratio may be 12 or more to one.

Q. Do thunderstorms occur in polar regions?

A. Yes, but not as often as in hotter places. Eskimos near the North Pole might see such a storm only once in two or three years, whereas people in Java can expect thunderstorms on more than 220 days each year.

Q. What is the air pressure at sea level?

A. It is 14.7 pounds per square inch. The pressure on the average person would be around 40,000 pounds.

Q. Was Seattle's site covered by a glacier during the last Ice Age?

A. Yes. This area probably was ice-covered for a thousand years or more. The so-called Fraser Ice Sheet extended into Washington, and its Puget Sound lobe reached maximum development about 14,000 years ago.

Q. Does our soil in Washington and Oregon acquire salt from rain?

A. Yes, a little. Ocean waves cause some sea salt to be hurled into the atmosphere. A cubic mile of air off the Pacific Ocean would contain anywhere from 10 to 100 pounds of salt, and some of the salt grains would become nuclei for raindrops.

Q. Where is the sea water coldest, and where warmest?

A. The coldest expanses of ocean are in the Arctic and Antarctic. The coldest temperature is 28.4 degrees, the freezing point of sea water of average salinity. The warmest ocean waters are in equatorial regions where the average temperature is 82 degrees and frequently reaches 86 degrees. Parts of the Red Sea—not an open ocean—get a few degrees warmer.

Index